Praise for

EINSTEIN'S GREATEST MISTAKE

Named "The Science Book of the Year" by the *Sunday Times* (UK)

Named one of the "Top Five Science Books of 2016" by ABC News (Australia)

"What Bodanis does brilliantly is to give us a feel for Einstein as a person. I don't think I've ever read a book that does this as well . . . Whenever there's a chance for storytelling, Bodanis triumphs."

—*Popular Science*

"Fascinating."

—*Forbes*

"Highly entertaining . . . For anyone interested in the greatest scientist of our time, this book should be required reading."

—*Astronomy Magazine*

"A thoroughly engaging and illuminating read."

—*Shelf Awareness*

"Wonderful . . . The best part is the real feel it gives of Einstein the man, and his thinking."

—*Cosmos*

"Bodanis has . . . [an] extraordinary ability to explain complicated physics . . . Bodanis is a lot like Einstein—minus the great mistake. Both see fun in physics, both love simplicity and brevity. In this book, theories of the universe morph into theories of life."

—*The Times* (UK)

"Bodanis makes Einstein's theories graspable . . . [His] biography offers a window onto Einstein's achievements and missteps, as well as his life."

—*BookPage*

"Engaging." —*Kirkus Reviews*

"Perceptive." —*Sunday Times* (UK)

"While we now remember Einstein for his early success and have reinvented him as a meme with crazy hair and sticking his tongue out, [*Einstein's Greatest Mistake*] reminds us to go beyond the cliché and remember the human—flawed, hubristic, and alone— but no less the greatest genius of the modern age."

—ABC News (Australia)

"I urge you to buy David Bodanis's new book *Einstein's Greatest Mistake* . . . it's a wonderful, fresh, and readable take on one of the most fascinating lives in science."

—Tim Harford, best-selling author of *The Undercover Economist*

"Einstein's greatest mistake was not his introduction of the Cosmological Constant in his famous equation of General Relativity, but rather his stubborn unwillingness to accept the reality of quantum mechanics. Once he divorced his thinking from the experimental forefront, his contributions to physics waned. David Bodanis gently describes Einstein's growing detachment from modern physics, shedding light on both Einstein the man and on several of his theories that did change our concept of the Universe."

—Lawrence M. Krauss, best-selling author of *The Physics of Star Trek* and *A Universe from Nothing*

"An intimate biography touching on the romances and rivalries of the celebrated physicist, as much as on his scientific goals, *Einstein's Greatest Mistake* reveals what we owe Einstein today—and how much more he might have achieved if not for his all-too-human flaws." —*BookReporter*

"Writer and futurist Bodanis imparts fresh insight into the genius—and failures—of the 20th century's most celebrated scientist . . . This provocative biography illuminates the human flaws that operate subtly in the shadows of scientific endeavor."

—*Publishers Weekly*

"Bodanis crafts an accessible introduction to this iconic genius."

—*Booklist*

Praise for DAVID BODANIS

"No one makes complex science more fascinating and accessible—and indeed more pleasurable—than David Bodanis."

—Bill Bryson, author of *A Short History of Nearly Everything* and *A Walk in the Woods*

"Richly original . . . Nothing short of astounding."

—*New York Times,* on *The Secret House*

"Astonishing . . . lucid, even thrilling . . . I didn't know I could know so much."

—Fay Weldon, Books of the Year, *Washington Post,* on $E=mc^2$

"Mesmerizing. A marvelously captivating book."

—*Washington Post Book World,* on *The Secret House*

"A technological odyssey complete with heroes and villains, triumph and tragedy—a true scientific adventure."

—Simon Singh, author of *Fermat's Enigma* and *Big Bang,* on *Electric Universe*

EINSTEIN'S GREATEST MISTAKE

EINSTEIN'S GREATEST MISTAKE

A Biography

DAVID BODANIS

Mariner Books
Houghton Mifflin Harcourt
BOSTON NEW YORK

First Mariner Books edition 2017

hmhco.com

Library of Congress Cataloging-in-Publication Data
Names: Bodanis, David, author.
Title: Einstein's greatest mistake : a biography / David Bodanis.
Description: Boston : Houghton Mifflin Harcourt, 2016.
Identifiers: LCCN 2016022348 (print) | LCCN 2016026595 (ebook)
ISBN 9780544808560 (hardcover) | ISBN 0544808568 (hardcover)
ISBN 9780544808584 (ebook) | ISBN 9781328745521 (pbk.)
Subjects: LCSH: Einstein, Albert, 1879–1955. | Physicists—Biography. Classification:
LCC QC16.E5 B66 2016 (print) | LCC QC16.E5 (ebook) | DDC 530.092 [B]—dc23
LC record available at https://lccn.loc.gov/2016022348

Book design by Martha Kennedy

Printed in the United States of America
DOC 10 9 8 7 6 5 4 3 2 1

To my son, Sam

Contents

Einstein walking home in Princeton, 1953

Prologue

PRINCETON, 1953. The tourists generally stayed on the sidewalk across the street from the white clapboard house on Mercer Street. But it was hard to keep down their excitement once they spotted the old man walking slowly back from the university campus, often wearing a long cloth coat and—if the New Jersey wind was especially sharp—a dark knit cap over his famous unruly hair.

The bravest tourists sometimes crossed over to say how much they admired him or to ask for his autograph. Most were tongue-tied or too awed to speak, and kept a respectful distance. For this old man was Albert Einstein, the greatest genius of all time, just yards away from them, his wise, wrinkled face suggesting he'd achieved insights deeper than other humans possibly could.

Einstein was the most famous scientist alive, but despite his celebrity he usually walked alone, or occasionally with one old friend. Although he was feted in public, and still constantly invited to white-tie dinners and even movie openings—Hollywood stars were especially excited to be photographed beside him—working scientists had little to do with him, nor had they for many years.

It wasn't his age that made them treat him this way. The great Danish physicist Niels Bohr was sixty-eight years old to Einstein's seventy-four, but Bohr remained so open to new ideas that bright doctoral students liked nothing more than spending time with him at his intellectually sparkling institute in Copenhagen. Einstein, however, had been isolated from mainstream research for decades. There was polite applause, of course, on the few occasions when he gave a seminar at the Institute for Advanced Study, in its forbidding plot on the edge of the Princeton campus, but it was the applause one might give an elderly soldier being wheeled out onto a stage. Einstein's peers regarded him as a has-been. Even many of his closest friends no longer took his ideas seriously.

Einstein could sense his isolation. At one time, his house would have been full of colleagues, youthful energy, the buzz of conversation. But lately it had become quiet. His second wife, the ever plumper and ever chattier Elsa, had passed away several years before, as had his beloved younger sister, Maja.

His sister's death especially pained Einstein. Maja and Albert had been constant companions as children back in 1880s Munich, teasing each other and building card castles. If a particularly elaborate castle collapsed in a gust of air, she remembered, her brother would doggedly start building it back up again. "I might not be more skilled than other scientists," he liked to say, "but I have the persistence of a mule."

Einstein had retained his youthful stubbornness, but his health wasn't what it had been. His main room, where he had his books and papers, was upstairs in his Princeton home, down the hall from the bedroom that had been Maja's. At his age, Einstein could climb the stairs only slowly, pausing for breath. But maybe it didn't matter. When he did settle in his study, he would have all the time in the world.

He was the greatest mind of the modern age. How had he ended up so alone?

WARTIME BERLIN, 1915. Einstein had just created a magnificent equation—not his famous $E=mc^2$ that had come ten years earlier, in 1905, but something even more powerful: the equation at the heart of what is called general relativity. It is one of the finest achievements of all time, as great as the works of Bach or Shakespeare. Einstein's 1915 equation had just two central terms, yet it would reveal unimagined features of space and time, explaining why black holes exist, showing how the universe began and how it will likely end, and even laying the foundation for revolutionary technologies such as GPS navigation. Einstein was overwhelmed by what he had discovered. "My boldest dreams have now come true," he wrote to his best friend that year.

But his dreams were soon interrupted. Two years on, in 1917, he realized that astronomical evidence about the shape of the universe seemed to contradict his general theory of relativity. Unable to account for the discrepancy, he dutifully modified his new equation, putting in an additional term that destroyed its simplicity.

As it turned out, the compromise was only temporary. Some years later, fresh evidence proved that his original and beautiful idea had been correct, and Einstein reinstated his original equation. He called his temporary modification "the greatest blunder of my life," for it had destroyed the beauty of his original, simple 1915 equation. Yet while that modification had been Einstein's first big mistake, his greatest error was still to come.

Einstein felt that he had been wrong to follow such faulty experimental evidence—that he should have simply held his nerve till the astronomers realized they had been mistaken. But from that he drew the additional conclusion that in the most important matters, he never had to follow experimental evidence again. When his critics tried to bring in evidence against his later beliefs, he ignored them, confident that he would be vindicated again.

This was a very human response, but it had catastrophic implications. It undermined more and more of what Einstein tried next, especially in the burgeoning study of the ultrasmall, of quantum mechanics. Friends such as Niels Bohr begged him to see reason. They knew that Einstein's exceptional intellect could transform the world yet again, if he only would let himself accept the new findings—valid ones—that a fresh generation of experimentalists were uncovering. But this Einstein could not do.

He had a few private moments of doubt but suppressed them. In his theory of 1915, he had revealed the underlying structure of our universe, and he had been right when everyone else had been wrong. He wasn't going to be misled again.

That conviction is what isolated him from the new generation's exciting work in quantum mechanics and destroyed his reputation among serious scientists; that is what left him so alone in his Mercer Street study.

How that happened—how genius reaches its peak and how it fades; how we deal with failure and with aging; how we lose the habit of trust and whether we can get it back—are the topics of this book. So, too, are Einstein's ideas themselves—right and wrong—and the steps by which he arrived at them. In that sense, this is a double biography: it's the story of a fallible genius, but also the story of his mistakes—how they began, grew, and locked in so deeply that even a man as wise as Einstein was unable to work himself free.

Genius and hubris, triumph and failure, can be inextricable. Einstein's 1915 equation, and the theory it undergirded, was perhaps the greatest feat of his life, yet it also sowed the seeds for his most astonishing failure. And to understand what Einstein did achieve in 1915, and how he went wrong, it's necessary to go even further back in time—to Einstein's earliest years, and the mysteries that intrigued him even then.

Part I

ORIGINS OF GENIUS

Einstein at university, around 1900

ONE

Victorian Childhood

TWO GREAT CONCEPTS dominated European science in 1879, the year of Einstein's birth, and they would provide the context for much of his greatest work. The first was the recognition that the forces that made the world's great industrial civilization function—the firing of coal in huge steam trains; the explosion of gunpowder in the warship turrets that kept subjugated peoples under control; even the faint pulses of electricity in the undersea cables that carried telegraphic messages around the world—were all but different manifestations of one fundamental entity, called energy. This was one of the central scientific ideas of the Victorian era.

Late Victorian scientists knew that energy behaves according to certain immutable principles. Miners could hack coal out of the ground, and technicians could feed gases from baking that coal into pressurized tubes that powered the streetlamps of London. But if something went wrong and the gas exploded, the energy of the resulting explosion—the energy of the flying shards of glass, plus the acoustic energy in the booming air and even any potential energy in errant fragments of metal from a streetlamp flung onto the rooftops nearby—would be exactly the same as the energy inherent in the gas itself. And if one fragment of streetlamp metal then fell to the pavement, the sound and energy of it hitting the ground, plus the energy of the gusts of wind as the fragment plummeted, would be exactly equal to the energy that had lifted it up in the first place.

The realization that energy cannot be created or destroyed, only transformed, seemed simple, but it contained extraordinary implications. When, for instance, one of Queen Victoria's servants opened the door of her carriage as it arrived at Buckingham Palace in central London, the energy that had been in his shoulder began to leave it . . . while exactly the same amount of energy appeared in the swinging motion of the ornate carriage door and the ever so slightly raised temperature of the friction-grinding hinge on which it turned. When the monarch stepped down to the ground, the kinetic energy that had existed in her descending form was transferred to the earth beneath her feet, leaving her stationary, but making our planet tremble in its orbit around the sun.

All types of energy are connected; all types of energy are neatly balanced. This simple truth became known as the law of the conservation of energy and was widely accepted by the mid-nineteenth century. Victorian confidence in religion had been bruised when Charles Darwin showed that a traditional God wasn't needed to create the living species on our planet. But this vision of an unchanging total energy was a consoling alternative. The way energy was so magically balanced seemed to be proof that some divine hand had touched our world and was still active among us.

By the time energy conservation was understood, Europe's scientists were well acquainted with the second great idea that dominated nineteenth-century physics: matter never entirely disappears either. In the Great Fire of London back in 1666, for example, Europe's largest city had been attacked by flames exploding from the tar and wood of the bakery where it began; roaring from one wooden housetop to another; pouring out vast volumes of acrid smoke; turning homes, offices, stables, and even plague-carrying rats into hot ash.

No one in the 1600s could have seen that as anything more than rampant chaos, but by 1800, a century before Einstein, scientists realized that if someone had been able to weigh absolutely everything in London before the flames began—all the wooden floorboards in all the houses; all the bricks and furnishings; all the beer kegs and even the scurrying rats—and then, with an even greater effort, had been able to measure all the smoke and ash and crumbling brick produced by the fire, it would come out that the weight of the two was, precisely, the same.

This principle became known as the conservation of matter and had been getting ever clearer from the late eighteenth century. Different terms have been used for this idea at different times, but the gist has always been the same: Burn wood in a fireplace, and you'll end up with ashes and smoke. But if you were somehow able to put a huge impermeable bag over the chimney and any drafty windows, and then you could measure all the smoke you captured plus all the ash—and take into account the oxygen pulled in from the air during the burning—you would find that the total weight was again exactly, precisely, the same as the weight of the firewood. Matter can change shape, turning from wood into ash, but in our universe it will never, ever disappear.

Those two ideas—the conservation of matter and the conservation of energy—would be central to the education and spectacular achievements of the young Einstein.

WHEN EINSTEIN HAD BEEN BORN, in 1879, in the German city of Ulm, some seventy-five miles from Munich, his family was just a few generations removed from the life of the medieval Jewish ghetto. To many Christian Germans of the nineteenth century, the Jews in their midst were strange, possibly subhuman, interlopers. To the Jews, however, virtually all of whom were Orthodox, it was the world outside their community that was threatening and disturbing, and never more so than when Christianity itself began to weaken, for that lowered the boundaries between the two religions. This let ideas of the eighteenth-century Enlightenment—

ideas about free inquiry, and science, and the belief that wisdom could come from studying the external universe—begin to enter, at first furtively, then ever more quickly, into the Jewish community.

By the generation of Einstein's parents, those ideas seem to have served Germany's Jews well. His father, Hermann, and uncle Jakob were largely self-taught electrical engineers, working on the latest technology of the time, creating motors and lighting systems. When Albert was an infant, in 1880, Hermann and Jakob moved together to Munich to set up a business in the uncle's name—Jakob Einstein & Co.—hoping to supply the city's growing electrical needs. Einstein's uncle was the more practical partner. Hermann, the father, was a dreamier sort, who had been fond of mathematics himself, but had had to leave school as a teenager to help in making a living.

Theirs was a warm family, and as Albert grew up his parents looked out for him. At around age four, Einstein was allowed to walk the streets of Munich on his own—or so his parents let him think. At least once, one of them—probably his mother, Pauline—followed him, well out of sight, but keeping an eye on how young Albert crossed the horse-busy roads to be sure that he was safe.

When Albert became old enough to understand, his father, uncle, and their regular houseguests explained to him how motors worked, how lightbulbs worked—and how the universe was divided into an energy part and a matter part. Albert soaked up these ideas, just as he assimilated his family's view that their Judaism was a heritage to be proud of, even if they felt that much of the Bible and the customs of the synagogue were little more than superstition. Leave that behind, they believed, and the modern world would accept them as good citizens.

By the time he was a teenager, though, Einstein recognized that Munich was an unwelcoming place, however much his family had tried to blend in. Back when he was six, his father's firm had secured a contract for the first electric lighting of the city's Oktoberfest. But as the years went on, contracts for the city's new lighting systems and generators went increasingly to non-Jewish firms, even if their products were inferior to those of the Einstein brothers. There were rumors that business prospects were better in prosperous Pavia, in northern Italy near Milan. In 1894 his parents and sister, Maja, moved there, along with his uncle, to try reestablishing the business. Albert, age fifteen, stayed behind, boarding with another family to finish high school.

It was not a happy time. The gentleness of the Einstein family was in sharp contrast to the harshness of the schools Albert attended. "The teachers . . . seemed to me like drill sergeants," Einstein reminisced decades later. They insisted on rote learning, aiming to produce terrified, obedient students. Famously, when Einstein was about fifteen and increasingly fed up with classes, his Greek teacher, Dr. Degenhart, had yelled, "Einstein, you'll never amount to anything!"—a comment that later prompted his ever loyal sister, who recorded the anecdote, to quip, "And indeed, Albert Einstein never did attain a professorship of Greek grammar."

Einstein dropped out of school when he was sixteen. If he had been forced out, he might have considered it a failure, but since it was his own choice, he actually felt proud, seeing it as an act of rebellion. He traveled on his own to join his family in Italy, worked for a while at his father and uncle's factory, and then reassured his worried parents that he had discovered a German-language university that didn't require a high school diploma and had no minimum age requirement. This was the Swiss Polytechnic in Zurich, and he applied right away. Although his math and physics grades were excellent—those family conversations hadn't been wasted—he should have paid more attention to Degenhart, for Einstein later remembered that he'd made no attempt to prepare, and his scores in French and chemistry let him down. The Swiss Polytechnic turned him away.

His parents weren't too surprised. "I got used a long time ago," his father wrote, "to receiving not-so-good grades along with very good ones." Einstein accepted that it had been a mistake to apply so early. He found a family to lodge with in the valleys of northern Switzerland near Zurich over the next year, as he took remedial classes to prepare for a second try.

Einstein's hosts in Switzerland, the Winteler family, assumed as a matter of course that he would sit around the table with them to share in reading aloud or discussion. They shared musical evenings —Einstein was a gifted violinist, whom school assessors had ranked highly back in Germany—and even better there was a daughter, Marie, who was just a bit older than him. Einstein seems to have thought it a token of affection to suggest that Marie do his laundry for him, as his mother had always done. He soon learned more sophisticated methods of courtship, however, and so began his first romance. This relationship triggered his mother's first bout of nosiness. When he was home with his family over the holidays and wrote to Marie, "Beloved sweetheart . . . you mean more to my soul than the whole world did before," his mother inked on the envelope the unpersuasive assertion that she hadn't read what was inside.

EINSTEIN MANAGED TO GET into the Polytechnic on his next try at age seventeen in 1896, on a course designed for the training of future high school teachers. He had just enough education to follow the lessons, yet enough of a cautious attitude from his already well-traveled life to judge them critically. It was the perfect background to cause him to take an independent view of what his teachers offered.

Although the Zurich Polytechnic was generally first-rate, a few professors were out-of-date, and Einstein managed to irritate them. Professor Heinrich Weber, for instance, who taught physics, had been helpful to Einstein at the beginning, but he turned out to have no interest in contemporary theory and refused to incorporate the Scotsman James Clerk Maxwell's groundbreaking work on the links between electrical and magnetic fields into his physics lectures. This irked Einstein, who recognized how important Maxwell's work could be. Weber, like many physicists of the 1890s, didn't feel there was anything fundamentally new to learn and believed that his job was simply to fill in remaining details. All the main work of figuring out the laws of the universe was complete, the thinking went, and although future generations of physicists might need to improve their measuring equipment so as to more accurately describe the known principles, there were no major insights left to be made.

Weber was also immensely pedantic, once making Einstein write out an entire research report for a second time, on the grounds that the first submission was not written on paper of exactly the proper size. Einstein mocked the professor by pointedly calling him *Herr* Weber instead of *Professor* Weber and harbored a grudge against him about his teaching style for years to come. "It is nothing short of a miracle that [our] modern methods of instruction have not yet entirely strangled the holy curiosity of inquiry," Einstein wrote about his university education a half century later.

Since there was little point in going to Weber's lectures, Einstein spent a lot of time getting to know the cafés and pubs of Zurich: sipping iced coffee, smoking his pipe, reading, and gossiping as the hours went by. He also found time to study, on his own, works of von Helmholtz, Boltzmann, and other masters of current physics. But his reading was unsystematic, and when the annual examinations came around, he realized he would need help catching up with Herr Weber's lesson plan.

What Einstein really needed was a fellow student to whom he could turn. His best friend was Michele Angelo Besso, a Jewish Italian who was a recent graduate of the Polytechnic, a few years older than Einstein. Besso was friendly and cultivated—he and Einstein had met at a musical evening where they were both playing the violin—but he had been almost as dreamy in class as Einstein had been. This meant Einstein needed to find someone else to borrow lecture notes from if he was to have any chance of passing, not least because one of his academic reports at the Polytechnic contained

Einstein's best friend, Michele Besso, 1898. "Einstein the eagle took Besso the sparrow under his wing," Besso once said in describing their intellectual partnership, "and the sparrow flew a little higher."

the ominous inked remark "director's reprimand for nondiligence in physics practicum."

Luckily, another of Einstein's acquaintances, Marcel Grossmann, was just the sort of individual every undisciplined undergraduate dreams of having as a friend. Like Einstein and Besso, Grossmann was Jewish and also only recently arrived in the country. Switzerland had a semiofficial policy of anti-Semitism at its universities that channeled Jews and other outsiders into what were then considered lower-status departments such as theoretical physics rather than fields such as engineering or applied physics, in which salaries were likely to be higher. (This wasn't too bad for Einstein, for it was only through theoretical physics that he was able to get a grip on concepts such as energy and matter that so intrigued him.) Knowing they were being treated in the same biased way probably helped Einstein and Grossmann bond.

When final exams came around, Grossmann's lecture notes—with all the important diagrams neatly drawn—did wonders for Einstein ("I would rather not speculate how I might have fared without them," Einstein wrote Grossmann's wife much later), enabling him to pass geometry, for example, with a respectable 4.25 out of 6. His score wasn't as good as Grossmann's, of course, which as everyone expected was a perfect 6.0. But none of his friends were surprised, for Einstein had yet another distraction.

Besides Besso and Grossmann, Einstein was spending time with another student, someone who was even more of an outsider than him: an Orthodox Christian Serbian, and the only woman in the course. With Mileva Marić's mix of high intelligence and darkly sensual looks, more than one student at the Polytechnic was interested in her. She was a few years older than the other students, was a skilled musician and painter, excellent at languages, and had studied medicine before switching to physics. Einstein had long since broken up with Marie Winteler from his lodging days and was ready to move on.

Grossmann and Einstein, several years after university, early 1910s

Mileva Marić, late 1890s. In 1900, Einstein wrote to her, "We shall be the happiest people on earth together, that's for sure."

Einstein was surprisingly handsome as a young man, with black curly hair and a confident, easy smile. His close relationship with his sister, Maja, had given him an ease with women and worked to his advantage when he began courting Marić. Over their undergraduate years, their romance advanced deeply. "Without you," he wrote to her in 1900, "I lack self confidence, pleasure in work, pleasure in living." But if they lived together, he promised her, "we shall be the happiest people on earth together, that's for sure." Throwing caution to the wind, at one point he had even sent her a letter with a drawing of his foot so that she could knit him some socks.

Einstein and Marić had held back for a while before telling their friends how close their romance had become, but they had been fooling no one. When Einstein was visiting his parents in Italy in 1900, he wrote to her, "Michele has already noticed that I like you, because . . . when I told him that I must now go to Zurich again, he asked: 'What else would draw him [back]?'" What else, indeed, but Marić?

There's something momentous about the years just before a new century begins, and Einstein's circle likely felt that sense of excitement. The four friends—Besso, Grossmann, Einstein, and Marić—had an attitude many students shared: the majority of their professors were relics from another age and not to be taken seriously, but the dawning twentieth century would bring forth wonders, and it would be the younger generation who would see them through to fruition. Of that, none of them seemed to have any doubt.

Each had their own source of confidence. Besso's family had a prosperous engineering business waiting for him in Italy, and he had already been spending time there as well as in Zurich. He was good with people, and confident that when he did eventually settle down he would be able to continue his family's success in industry. Grossmann had a stand-out mathematical talent that everyone at the Polytechnic recognized. Marić had been a superb student at her technical secondary school in Budapest and in fact had been one of the first women in the Austro-Hungarian Empire to attend a scientific high school at all. She was also one of the few female university students in Switzerland. In a country where woman suffrage was still seven decades away, this was even more of a distinction for her.

The four friends were hungry to advance the world's knowledge—Einstein perhaps most of all. Although he struggled with his schoolwork, his private intellectual pursuits were picking up steam. Along with those long hours reading newspapers and playing the joker in Zurich's cafés, he had continued to study Europe's greatest physicists, teaching himself everything the out-of-date Professor Weber ignored.

Einstein was fascinated by the ideas of Michael Faraday and James Clerk Maxwell that there could be invisible fields of mixed electricity and magnetism stretching through space, influencing everything within their reach. He was fascinated by more recent findings as well: J. J. Thomson at Cambridge measuring details of the electron, a tiny particle that seemed to exist inside the atoms within every substance; Wilhelm Röntgen discovering X-rays that could see through living flesh; Guglielmo Marconi sending radio signals across the English Channel. How did these phenomena occur, Einstein wondered, and why? He had been mulling this over since the year he'd spent in Italy with his family before going to Switzerland, but he had been unable to take his inquiries any further then.

Now he was eager to advance not just his own knowledge but also that of the entire field of physics. Einstein owed part of his newfound drive to his desire to help out his father, whose companies in Pavia and Milan, despite the relative lack of anti-Semitism there, were no more successful than his previous partnership in Munich. The money his parents sent him to live on was a great drain on them, and he knew it. Einstein also owed part of his drive to what he had drawn from his religious heritage. Although he had dropped the formalities of religion at the age of twelve, he did believe there were truths waiting to be found in the universe, only some of which mankind had glimpsed. That would be his quest, he vowed in an 1897 letter to Marie Winteler's mother.

"Strenuous intellectual work," he wrote, "and looking at God's Nature are the . . . angels that shall lead me through all of life's troubles . . . And yet what a peculiar way this is . . . One creates a small little world for oneself, and as lamentably insignificant as it may be in comparison with the perpetually changing size of real existence, one feels miraculously great and important."

For most of Einstein's friends, those feelings of general grandeur to come were as far as they went. He, however, was thinking about the great Victorian synthesis a lot now and beginning to question the grand vision that had been handed down to him. The universe was divided into two great realms. There was energy, as carried in the gusting winds that blew down the Zurich streets he knew so well, and there was matter, such as the glass windows of his beloved cafés and the beer or mocha he sipped as he thought about all these things. But did the unity have to stop there?

At this stage, young Einstein couldn't do anything more with such a thought. He was intelligent, but the questions he was asking himself seemed impossible to answer. And he was young enough to simply make do with the dominant vision of the universe as having two unlinked parts—albeit with the confidence that he could come back to it later.

TWO

Coming of Age

UNIVERSITY FRIENDS LIKE to imagine they'll stay together forever, but it rarely turns out that way. In 1900, Einstein's, Grossmann's, and Marić's four years at the Zurich Polytechnic were up. Besso, a few years older, had already moved back to Italy to work in electrical engineering. Although Einstein tried to argue him out of it ("What a waste of his truly outstanding intelligence," he wrote Marić that year), he respected Besso's decision, which would prevent him from becoming a financial burden to his family. Grossmann was going to teach high school, though he had his eye on research and eventually registered for graduate study in realms of pure mathematics that perplexed the more practical Einstein. Marić was caught between staying in Switzerland for more studies (and her boyfriend) and going back to her family near Belgrade, whom she now had to visit.

Einstein also was stuck. He badly wanted to pursue a career as a research scientist, but he had so upset his main physics lecturer, Professor Weber, through his insubordination and skipping classes that Weber now refused to write letters of recommendation to other professors or school heads, which was the usual way for students to procure such jobs after graduation. With remarkable aplomb, Einstein himself tried writing to one of his former mathematics instructors, Professor Hurwitz, explaining that although he had, in fact, not bothered to attend most of Hurwitz's classes, he was writing "with the humble inquiry" as to whether he might be granted a job working as Hurwitz's assistant. For some reason, Hurwitz was not impressed, and there was no job there either. Einstein kept on writing letters—"I will soon have graced every physicist from the North Sea to the southern tip of Italy with my offer," he wrote Marić—but all he got were rejections.

These especially hurt because he knew his family needed more income. A little earlier, he'd told Maja, "What oppresses me most, of course, is the [financial] misfortune of our poor parents. It grieves me deeply that I, a grown man, have to stand idly by, unable to do the least thing to help."

After a stint as a high school teacher, and even for a while working as a tutor to a young Englishman in Switzerland, Einstein was back living with his parents in Italy in 1901. His father, Hermann, recognized that his son was depressed and resolved to help. He decided to write to Wilhelm Ostwald, one of Germany's greatest scientists, explaining that "my son Albert is 22 years old [and] . . . feels profoundly unhappy . . . His idea that he has gone off the tracks with his career & is now out of touch gets more and more entrenched each day." Hermann asked the professor to write Albert "a few words of encouragement, so that he might recover his joy. If, in addition, you could secure him an Assistant's position for now or the next autumn, my gratitude would know no bounds." Naturally, this had to remain between the two of them, for "my son does not know anything about my unusual step." The appeal was heartfelt, but it rambled and was as ineffectual as most of Hermann's business ventures. Ostwald never replied.

As for Einstein's relationship with Marić, although his mother hadn't met her, she couldn't bear this girl he spoke of so much—for what woman, if one thought about it, was ever going to be good enough for her son? Pauline used Einstein's failure to earn a good living as a further reason to insist that he stop writing to this non-Jewish woman. After three weeks of this moral torture, Einstein wrote to Grossmann in desperation, asking if there was any way Grossmann could help him escape having to live at home. When Grossmann called on family connections, securing Einstein an interview at the Patent Office in Bern, Einstein immediately wrote back, "When I found your letter I was deeply moved [that] you did not forget your old luckless friend."

It wasn't exactly the profession Einstein had envisioned for himself, but the Patent Office job—if he could get it—would be a useful way to make a living, and just maybe protect his relationship with Marić from his mother as well. It helped that earlier in 1901, Einstein had taken out Swiss citizenship, having passed an application process that involved being shadowed by a private detective, who noted that Herr Einstein kept regular hours, scarcely drank, and so deserved to be approved. But even so, the position seemed a letdown, merely a way to earn a steady paycheck while he tried to get back into the academic system. He had to pretend to his parents that this was fine and no stumbling block at all.

At least everything was continuing to go well with Marić, for while he was still with his parents in northern Italy, she was back in Switzerland, not too terribly far away. They could write to each other about science and love—and they could arrange to meet.

May, 1901

> My dear dolly! . . . This evening I sat 2 hours at the window and thought about how the law of interaction of molecular forces could be determined. I've got a very good idea. I'll tell you about it on Sunday . . .
>
> Ah, writing is stupid. Sunday I am going to kiss you in person. To a happy reunion! Greeting and hugs from your,
>
> Albert
>
> PS: Love!

And kiss they did, finally meeting in the Swiss Alps, high above Lake Como. Marić wrote to her best friend describing how she and her boyfriend had to cross a pass in twenty feet of snow.

> We rented a very small [horse-drawn] sledge, the kinds they are using there, which has just enough room for 2 people in love with each other, and the coachman stands on a little plank in the rear . . . and calls you "signora"—could you think of anything more beautiful?
>
> . . . There was nothing but snow and more snow as far as the eye could see . . . I held my sweetheart firmly in my arms under the coats.

Einstein must have held her just as firmly. "How beautiful it was," he wrote her, "[when] you let me press your dear little person against me, in that most natural way." By the end of their holiday in May 1901, she was pregnant. Given the mores of the time, Marić had no option when she found out but to return to her family until the birth. Nine months later, Einstein wrote her.

> Bern Tuesday [February 4, 1902]
>
> It has really turned out to be a little girl, as you wished! Is she healthy and does she cry properly? What kind of little eyes does she have? Is she hungry?
>
> I love her so much & I don't even know her yet!

Few more references to their daughter survive, for it was nearly impossible for an unmarried couple from their backgrounds to keep an illegitimate child. Although they named their daughter Lieserl (Elizabeth), the indirect evidence suggests that they gave her up for adoption, probably to a family friend in Budapest. Einstein never spoke of her again.

AFTER A SERIES of interviews, Einstein did secure the Patent Office job for which his friend Grossmann's father had put in a good word. It was in the much smaller city of Bern—not Zurich, but still an acceptable location, even though the salary wasn't what Einstein had hoped for. He had applied for the position of Technical Expert Second Class, but the head of the Patent Office, Superintendent Haller, disappointed at Einstein's lack of technical capacity, had only offered him the lower-paid position of Technical Expert Third Class.

Einstein accepted the post, but he needed more money. Like his father, he was entrepreneurial, and in 1902 he put an advertisement in the local paper:

Private lessons in
MATHEMATICS AND PHYSICS
for students and pupils
given most thoroughly by
ALBERT EINSTEIN, holder of the fed.
polyt. teacher's diploma
GERECHTIGKEITSGASSE 32, 1ST FLOOR
Trial lessons free.

But if Einstein was just as energetic as his father, the two men also shared a certain vagueness about business details. Although he did attract several students, he was so pleasant and talkative that he became friends with most of them—and then felt he couldn't charge them for lessons. Somehow, however, he did gradually accumulate some savings, including from one student he continued to charge, and who has left a pen portrait of Einstein at this time: His tutor, he wrote, "is 5 ft 9, broad shoulders . . . large sensual mouth . . . The voice is . . . like the tone of a cello."

Einstein was also trying to continue his own research, but it was difficult. The Patent Office was a six-day-a-week job, and the one good research library in Bern was closed on Sunday, his sole day off. He was too proud to let anyone know how difficult his life was, and certainly too proud to apologize to Professor Weber and grovel his way back into academia.

Einstein may have been struggling professionally, but his romantic life was all he had dreamed of. Marić had some savings from her family, and with their combined money they could afford an apartment big enough for them both. She moved back to Switzerland, and in January 1903 they were married at the Bern City Hall. He was nearly twenty-four, and she was twenty-eight. They wouldn't have been human if they didn't miss their daughter. "We shall remain students together for as long as we live," Einstein wrote exultantly, "and not give a damn about the world."

His mother was still angry about his choice, letting everyone—and especially her son—know how much she hated Miss Marić. But his loyal younger sister, Maja, urged her to give Einstein's wife a chance. Marić herself was confident that she would ultimately win the Einstein family over: as she told a girlfriend, she'd simply find someone the mother respected, and make herself helpful to that person, and well, the mother would have to see how well-meaning she was then, wouldn't she?

The happy couple made new friends in Bern, helped by the fact that skilled violinists were always appreciated. Einstein was often invited to the homes of families that wanted an extra instrument for their own musical evenings. He and Marić also continued seeing the ever loyal, easygoing Michele Besso, who soon moved back to Switzerland from Italy and took a job at the Patent Office as well. Einstein told him, "So I'm a married man now . . . [Mileva] looks after everything splendidly, is a good cook, and is always cheerful." Besso was already married, too, and Einstein had played a part—introducing him to his ex-girlfriend Marie's family, which Besso enjoyed so much that he proposed to Marie's older sister Anna, and soon had a son with her. The couples easily spent time together. "I like him a great deal," Einstein wrote about Besso, "because of his sharp mind and his simplicity. I also like Anna, and especially their little kid." By the end of 1903, Einstein and Marić had moved into an apartment with a small balcony overlooking the Alps. They would squeeze onto the balcony—sometimes with their friends, sometimes just the two them—newlyweds admiring their luck.

• • •

EVER SINCE HIS teenage years, Einstein had had moments when he felt greatly isolated. Even now, surrounded by those he loved, he was conscious of the barriers that could separate people from one another, even if they'd been close or lived in the same home. He confided to Marić that he and his sister had "become so incomprehensible to each other that we are unable to . . . feel what moves the other," and that at times "everyone else seems alien to me, as if held back by an invisible wall." It would have seemed a small miracle that Marić herself had broken through.

When their first legitimate child—a son, Hans Albert—was born in 1904, their income was still low. ("When I talked about experiments with clocks at different parts of a train," Einstein remembered later about work he was soon to commence, "I still only possessed one clock!") But the young family had everything it needed. Einstein was good with his hands, and instead of buying his son expensive toys, he improvised with everyday items, once constructing an entire working miniature cable car set out of matchboxes and string, a memory his son cherished even decades later.

It was a happy time. The love between Einstein and Marić had survived the adoption of their daughter, professional frustrations, and the specter of poverty. Surely it could survive anything.

THREE

Annus Mirabilis

IT WAS AT the Patent Office in 1905 that Einstein had his first great breakthroughs.

In many respects, the office was as formal and constrictive as he had feared. It was part of the Swiss federal civil service, and there were strict hierarchies of rank. Einstein was just one of several dozen trained men working at nearly identical high desks through long, constantly supervised days.

Yet it was surprisingly interesting work and had a number of advantages for Einstein in his dream of getting back into the academic world. For one thing, at the Patent Office he was supposed to judge applications for new devices, especially in the field of electrical engineering, and decide if they were original enough to deserve a patent. This was a bit like getting an early look at the latest high-tech creations in Silicon Valley today, and many of the principles he developed for judging those applications would be useful in his later work.

Another upside of the job was the freedom it afforded him to pursue extracurricular work. Although his supervisor, Herr Haller, was pedantic, he tolerated the fact that Einstein obviously was spending free moments on his own research papers, which he would hurriedly push aside or cram into a desk drawer (which he cheekily dubbed his "Department of Theoretical Physics") whenever Haller strode near.

Since Einstein knew that his only chance of getting a university post would be by coming up with strong research findings, he felt none of the pressure to publish preliminary, incomplete findings he would have faced had he already obtained a university job and was working his way up ("a temptation to superficiality," he later wrote, "which only strong characters can resist"). If the task was daunting, he still wasn't going to let anyone else know just how formidable it was—aside, perhaps, from his wife, who had professional frustrations of her own. Marić had seen her own dreams of research crushed, having failed to obtain an academic post, and was now cooped up at home with their son. It would have been only natural for the two lovers to commiserate, even if the disparate causes of their suffering were slowly opening a gulf between them.

In the evenings, Einstein would go out for long walks with Besso and others, including a new friend named Maurice Solovine, a young Romanian who had applied for the physics lessons Einstein was still offering and had since become one of his crowd—even though he'd given up on physics after a session or two with Einstein and switched to philosophy instead. Sometimes Marić would join them; sometimes it would be just the men. They'd stop at country pubs for cheese, or beer, or the mocha that Einstein favored, and would talk about health foods or the newfangled "aerobics" exercise classes that were constantly being publicized, or about politics and philosophy and all their dreams for the future.

In the summer, if they'd been talking till very late, Einstein and his friends would continue to a mountain just outside Bern where the Einsteins also sometimes went in the daytime with Besso's family. "The sight of the twinkling stars," Solovine wrote, "made a strong impression on us." There they'd wait and, Solovine went on, "marvel at the sun as it came slowly towards the horizon, and finally appeared in all of its splendor to bathe the Alps in a mystic rose."

Physics, and the foundations of how the world was put together, were natural topics for such moments. Everything in Einstein's field had been accelerating since his graduation year at the Polytechnic. Marconi had now sent radio waves across not just the English Channel, but across the Atlantic. Marie Curie in Paris had found immense, seemingly limitless sources of energy in radium ores; Max Planck in Germany seemed to have shown that energy didn't pour out of gradually heated objects in a smooth way, but "jumped" in strange, abrupt intervals—which later became known as quantum jumps. Thermodynamics was a matter of great wonder, for how did the universe know to move heat around in the precise way it did? And there still was the odd way that everything fit into two seemingly perfectly balanced realms—the realm of energy and the realm of matter, or what scientists were increasingly thinking of as the realm of mass.* There had to be some simple unity behind it all, Einstein and Solovine and their closest friends believed: a handful of deep principles that would explain why the universe had been put together to make everything work.

But what?

After the long walks, and reflection at the mountains, there'd be a quick coffee at the nearest café, then a walk together back to town, where the wanderers would start their respective workdays. "We

---

* The terms used by early scientists had subtly different meanings from what they have now. To Lavoisier and others in the late eighteenth century, it was natural to think in terms of matter, what we today would think of as the number of atoms in an object. Gradually that changed, and by the early twentieth century the concept was understood in terms of the conservation of mass. What's the difference? "Mass" is most easily thought of as the measure of an object's resistance to acceleration. A pen is easily accelerated, a big mountain isn't, and so the latter has more mass. The twist is that the two different views are closely related: mountains are hard to accelerate not least because they have more atoms inside.

were overflowing with good spirits," Solovine recalled. There was no need for sleep.

The only problem was that Einstein wasn't, yet, quite as confident as he appeared. He knew how his father had never achieved what he'd hoped for, with one business venture after another not quite succeeding, leaving Einstein's parents always dependent on help from richer relatives. And he had seen his closest friends abandon their own lofty dreams for a chance at stability. Marić had put her research to the side, because of the birth and then abandonment of Lieserl, and Besso had slid away from research, too, first by returning to his family's engineering firm, and then by joining Einstein at the Patent Office.

Although Einstein's and Besso's day jobs were interesting, it was not the creative work they had once dreamed of. Einstein knew that the great Englishman Sir Isaac Newton had been only in his mid-twenties when, in the 1660s, he not only came up with the ideas for calculus but also had the first glimmers—at his mother's Lincolnshire farm, with the famously falling apple—of his great idea that a single law of gravitation stretched from inside the earth, up to apple trees on meadows above, and on to the moon itself, hurtling in its orbit a quarter of a million miles beyond. Einstein was the same age. Where was his great discovery?

Was Einstein going to be one of those who spent their entire lives on the sidelines, admiring what others achieved? To his little sister, Maja, he was a genius—the big brother who could do anything. But Einstein himself could have been forgiven for taking a gloomier view. In his spare time, he tried to put together ideas for publication, but as he turned twenty-four and then twenty-five, none of them were what he had hoped for; none were very deep. He examined the forces that help make liquids curve upward within a straw but didn't come up with anything profoundly original. If he hadn't become "Einstein," these papers would have been forgotten.

Time ticked by, and then as he approached twenty-six, something remarkable happened. In a flurry of activity in the spring of 1905, his deadlock broke, and Einstein began to write a series of five papers that very shortly would transform physics.

EINSTEIN'S MIND WAS pulling him in many different directions around the time of his twenty-sixth birthday. He was thinking about space and time, light and particles, and began drafting papers about these subjects. But as he did so, he also found himself returning to his earlier speculations about whether there might not be some deeper unity to the universe than he had been taught.

Einstein's upbringing may not have given him the keenest sense for business, but it had prepared him perfectly for this sort of intellectual freshness. As the Norwegian American economist Thorstein Veblen once observed, when families are undergoing a shift from religious belief to secularism, the children often grow up to be skeptical of any claims of ultimate truth—whether coming from authorities in religion, or science, or any other field. Einstein was shaped by that skepticism, as were other members of his family—especially his sister, Maja, whose unconventional perspective on things manifested itself in a keen sense of irony. (Recalling years later how Albert had once thrown a heavy ball at her head in a fit of temper, she noted, "This should suffice to show that it takes a sound skull to be the sister of an intellectual.")

Maja's skeptical attitude came out in teasing wit, but for Einstein this same trait led him to question everything he was taught, whether in his Munich secondary school, his Zurich Polytechnic, or in his own readings. And his inherent skepticism had been building toward this moment, although his fighting spirit would come in handy as well.

As his magnificent work in 1905 went on, Einstein began to investigate in earnest whether the two realms that his Victorian predecessors had believed were entirely separate were not actually linked in some way. The dominant view at the time, as his father and uncle and family friends had explained to him during his childhood, and as he and everyone in his classes in Zurich had drilled into them even further, was that the universe was divided into two parts. There was the domain of energy, which scientists referred to with the letter *E*. And there was the domain of matter—or, more technically, the domain of mass—which was symbolized by the letter *M*.

To scientists before Einstein, it was as if the entire world were divided into two vast domed cities. Inside the domed city of *E*, where energy existed, there were flickering flames, roaring winds, and the like. The other domed city, separate and located far away, was the land of *M*, of mass, where mountains and locomotives and all the other heavy, substantial stuff of our world existed.

Einstein became convinced that there had to be a way of uniting them. The God he didn't quite believe in had had no reason arbitrarily to stop creating the universe when He had reached two parts. If there was any sense, He would have gone further and created a deeper unity, of which everything we see is just a different manifestation.

Science is often described as depopulating the heavens—ridding them of mystical forces and beings, giving us a world in which cold reason is enough to account for all that we see. But Einstein was a student of the history of science, and he knew that he wasn't alone in feeling that there was something more. Newton, too, had written suggestively that he was simply seeing God's intentions in the laws he uncovered.

Newton's life had straddled the seventeenth and eighteenth centuries, and he saw no distinction between his research in what we now term physics and his research in what we today view as the separate fields of theology and biblical history. He believed the Bible contained hidden truths laid down by God, and that helped him believe the universe also held the Creator's secrets.

Over time, for most scientists, Newton's religious assumptions had come to be considered a vestige of the earliest phase of science—a scaffolding that might have been needed at the start but that, with maturity, could be taken away, allowing the "machine" of scientific investigation to operate on its own. The notion of a clockwork universe had begun to take over: of a universe that had intricately linked inner parts and might have been wound up once at the beginning by God, but that since then was able to advance quite automatically, on its own, with any need for a divine hypothesis or presence fading further and further into the past. Researchers in the eighteenth and especially the nineteenth centuries who felt otherwise were thought to have been fed archaic ideas when young, engaging in a touching homage to their community perhaps, but their beliefs otherwise of no significance.

Einstein didn't go along with that. For scientists at the highest level, he once said, science surpassed and replaced religion: "[Their] religious feeling takes the form of a rapturous amazement at the harmony of natural law, which reveals an intelligence of such superiority that, compared with it, all the systematic thinking and acting of human beings is an utterly insignificant reflection." Whoever did not have this sense of wonder "is as good as dead, and his eyes are dimmed." Newton had shown that our universe is organized by laws as succinct as the divine instructions he found within the Bible. Einstein, now age twenty-six, was ready to do the same.

So what if the universe wasn't really divided into two distinct parts after all? What if—to use the image provided earlier—the two domed cities weren't sitting in utter isolation on separate parts of a vast continent, but actually had a secret tunnel between them through which whatever was in one could hurtle along and take up shape in the other? What if everything in the city of *E*, of energy, could shoot through the tunnel and turn into mass, or *M*—and everything in the domed city of *M*, of mass, could shoot back through the tunnel and turn into energy, or *E*?

Imagining a tunnel between these domed cities is a bit like imagining that the energetic fire that flickered over a burning log wasn't different in nature from the material wood of that log, but rather, in some way, the wood could explode apart into flame, or—in the other direction—the fire could be squeezed back into the wood. In shorthand, that would be like saying that energy could become mass, and mass could become energy. Or, even more briefly, that *E* could become *M*, and *M* could become *E*.

THE POSSIBILITY THAT energy and mass were one and the same wasn't yet clear to Einstein. Yet as he was wrapping up his other work, in the summer of 1905, he realized that he could go further.

The vision of energy and mass as interlinked—of a tunnel connecting the city of *M* with the city of *E*—was at the heart of the final paper in Einstein's 1905 series. The question that he had to answer before he could publish this radical theory was how the tunnel between *M* and *E* operates in our real world. Does it transfer items back and forth directly, or does it somehow enlarge them when they travel in one direction and shrink them when they travel in the other? In the first case, it would be as if the world had only two cities—say, Munich and Edinburgh—and an invisible tunnel whooshed people back and forth between them without changing their size: just letting them arrive with a curious ability to speak the local language. In the second case, it would be as if each city's residents changed size when they arrived in the other city, a bit like Alice. But which city's citizens would shrink as they traveled, and which would grow?

Einstein worked this out in the late summer of 1905. He showed that the universe was arranged so that it was the objects in the "mass" city that would automatically seem to expand as they transformed into energy. In our example of Munich and Edinburgh, the pudgy mass burghers of Munich would enter the transformation tunnel as commuters of ordinary height, but then, when they'd finished their remarkable journey into Edinburgh, they would emerge from the tunnel as totteringly huge energy beings, hundreds of feet tall, able to bestride the city like enormous walking skyscrapers. In the other direction, when Edinburghians hurtled down the tunnel to Munich, they would shrink, and diminish, so much so that when the bewildered wee things emerged in Munich, they would be smaller than the tiniest fragments of mass-dense bratwurst that they saw street-corner vendors sold.

How much did each side change as it transformed? In solving this problem, Einstein brought in an entirely new approach that had come to him in that wondrous year—an idea that was as unexpected as a brilliant move in chess. We're used to thinking that if we're in a parked car and turn the headlights on, the light rays that shoot forward will be traveling at a certain speed, and then if we start driving and reach 60 mph, the light rays will now be traveling 60 mph faster as well. From deep principles, however, Einstein had concluded that this wasn't the case, and with further ingenious twists, he now managed to show that energy and mass transform into each other: that they're just different labels for what's actually one thing.

By this period, scientists had long known that the speed of light was very great. It's just a bit over 670,000,000 mph: enough for a signal flashed from earth to reach the moon in under two seconds or to cross the entire solar system in just hours. This speed—approximately 670,000,000 mph—was symbolized by the letter c, from the Latin *celeritas,* as in the English word "celerity," meaning "speed."

If mass were simply magnified by the factor of the speed of light as it traveled down the transformation tunnel, it would produce a tremendous amount of energy. But Einstein's calculations showed even that wasn't as far as matters went. Multiply c by itself, and one creates the even larger number c^2: approximately 450,000,000,000,000,000 mph^2. *That's* how much any bit of mass will be magnified when it's transformed into energy. Mass can become energy, in incredibly vast amounts. The large number c^2 says exactly what the change is. In shorthand, $E=mc^2$.

Most of the time, the energy inherent in mass stays hidden, since almost all substances on earth are very stable. Einstein often described the energy inside ordinary rocks or metals as like a huge pile of coins kept by a vastly wealthy miser: able to create great effects if they were let out, but invariably kept guarded within, and thus invisible to the outside world. But even in 1905, some experts were finding ways to let little bits out.

In Paris, Marie and Pierre Curie had become famous for experiments in which they'd found that radiant heat—a form of energy—would spray out of mere specks of radium ore: hour after hour, day after day, year after year. Today we realize that all that glowing energy was coming from a very few atoms transforming, multiplying by that factor of 450,000,000,000,000,000 as it sped outward and produced heat. Einstein knew of the Curies' work, and at the end of his final 1905 paper—still modest enough to know that any great idea requires some proof—he suggested, "Perhaps it will prove possible to test this theory using bodies whose energy content is variable to a high degree, e.g., salts of radium."

As summer turned to fall and Einstein put the finishing touches on his fifth and final paper and sent it off to the German journal *Annalen der Physik (Annals of Physics)*, he had no idea what lay ahead. Just forty years later, a great nation would configure purified uranium in such a way that entire ounces of that metal could be made to transform in accord with his equation—each fraction of mass becoming enhanced by the huge multiplier c^2 as it "disappeared" from material existence and instantly revealed itself as pure energy instead. The result, over Hiroshima, was a rush of energy exploding outward that destroyed an entire city: creating fires, hurricane-force winds, and a light flash so staggeringly intense that it hit the moon before reflecting back to earth. When Einstein, in exile in America, heard the news over the radio in 1945, he turned to his longtime secretary and said, distraught, that if he had known what was going to happen, he wouldn't have lifted a finger to help.

All that was far in the future. For now, the young physicist was satisfied with his work. His penultimate paper to the *Annalen der Physik* had been the one showing the central role that the speed of light played in a vast range of concepts. The work in that paper, published in September 1905, is what became known as special relativity. The day after it was published, the *Annalen* received his final paper, showing one particular consequence of that: the fact that mass and energy can be transformed into each other. This spin-off from special relativity was published on November 21, 1905, and completed his *annus mirabilis*—a most extraordinary year both for Einstein and for the world.

In just several months, the unknown Einstein had published several of the most significant papers in the history of science. He had seen how clearly the universe's inner operations were arranged, as with that hitherto unimagined tunnel between mass and energy that $E=mc^2$ so accurately described. These and the other concepts in his 1905 series would gradually revamp our understanding of everything from the operations of light to the nature of space and time. As physicists came to understand his work, they would also give its author a taste of the respect from his colleagues he so wished. Yet as the last of his papers was published in the fall of 1905, Einstein could only have guessed what lay ahead—and how much further he had yet to go.

Einstein was growing in confidence, but was still far from smug. When he first came up with the idea for his final paper, linking *E* and *M*, he had written a friend, "The idea is amusing and enticing, but whether the good Lord is laughing at me and leading me up the garden path—that I cannot know."

He was also exhausted from the months of intense labor. He had accomplished all this while still working six days a week, eight hours a day, at the Patent Office. When he was finally done, he and Mileva went out drinking, which was rare for them: Einstein rarely venturing beyond the occasional beer, and both of them generally had just tea or coffee at the table. Their lack of experience shows, for a postcard survives from the next day, signed by them both: "Both of us, alas, dead drunk under the table."

FOUR

Only the Beginning

IN THE SUMMER OF 1907, Max von Laue, a personal assistant of the great Berlin physicist Max Planck, was sent to Bern on a mission to meet the man who had published those extraordinary papers in the respected *Annalen der Physik* back in 1905.

When von Laue arrived and made his inquiries, he found that the man who he presumed would be a Herr Doktor Professor Einstein was not at the University of Bern, but seemed to be in residence in the post office building, which housed the Patent Office. Von Laue walked there and asked for the Professor to be called. Several minutes later, a polite young man walked through the waiting room. Von Laue ignored him, waiting for the Professor. The young man seemed confused—why had he been called if there was no one to greet him?—before returning to his desk up on the third floor.

Another request was put in: surely the Professor was not taking this long to come down? After von Laue waited a while longer, Einstein entered for the second time. Only then did Planck's assistant realize that this must be the great thinker: not a professor—not even a *Doktor*—but somehow a mere minor functionary in the post office building.

Maja remembered that Einstein thought his publication in the

Annalen would be immediately noticed and was disappointed when it seemed to be wholly ignored. Partly this was because he hadn't bothered to write up his results in the usual scientific form, with a multitude of footnotes referring to previous work by famous professors. There were few footnotes in his main paper, yet in the final paragraph he warmly thanked his friend Michele Besso, who had helped him through thoughtful back-and-forth discussions about physics as they'd taken long walks outside Bern. But partly it was because Einstein's achievement was hard to grasp.

Einstein had arrived at his theories using very general principles. This technique had served him well at the Patent Office, where he had learned how to use such higher-level principles to judge whether an invention was going to work or not. If an inventor said that a device sent in for assessment used perpetual motion, for example, Einstein knew he could reject the application right away. Perpetual motion isn't possible, not in our earthly world of friction and entropy. When applied to more ambitious projects, however, Einstein's simple, abstract approach often made his theories difficult for his scientific peers to wrap their heads around, let alone engage with.

In his 1905 works, Einstein had used a range of such higher-order principles to come up with ideas of shocking strangeness. There was $E=mc^2$ from his November paper, which insisted—quite accurately—that energy was just a very diffuse form of mass, and mass was simply exceptionally dense energy. For anyone schooled in mainline Victorian science, that contention was shocking enough. But that equation was just one consequence of the broader special relativity theory from his earlier, September paper—a theory that fundamentally reworked what it means to observe events in space and time.

Special relativity had other, equally bizarre implications besides those that Einstein would tease out in $E=mc^2$. If we watched a train traveling sufficiently fast, Einstein showed in that September paper, we would see it get shorter in the direction in which it was moving. Moving fast enough, the very largest locomotive would end up no thicker than a postage stamp. Time wasn't what we thought it was either. We're used to thinking that time always "flows" at the same rate for everyone. But someone accelerating at high speed away from the earth would see our entire species whirring through centuries in what seemed bare minutes, while we on earth, if we could watch the traveler on the spaceship through sufficiently powerful telescopes, would see his life slooooow almost to a stop. Both an observer on earth and the traveler would feel it was his own life that was normal and the other that had changed.

Could something this odd really take place? Many physicists—at least those who bothered to study Einstein's theory at all—objected to the notion at first. Theoretical physics was still a very small academic field, and one of its few professors, the distinguished Arnold Sommerfeld in Munich, wrote confidentially to a friend, "This unconstruable and unvisualizable dogmatism [of Einstein's] seems to me to contain something almost unhealthy. An Englishman would scarcely have produced this theory; perhaps it reflects . . . the abstract-conceptual character of the Semite."

Yet even Sommerfeld, when he worked through Einstein's reasoning, saw that it was irrefutable. We don't notice these strange consequences because they tend only to be visible at extremely high velocities, or in the rare cases when atoms are so fragilely constructed that they can fly apart, as with the radium samples that so perplexed Marie Curie. But if we ever entered into those realms, we would see that all the strange activities Einstein described were true.

Physicists may have been slowly coming around to Einstein's thinking by mid-1907, roughly a year and a half after the last of his annus mirabilis papers was published, but von Laue was the first major scientist to visit Bern. Einstein seized the opportunity not just to rub shoulders with the scientific elite, but also to see whether, in doing so, he might find a way to get himself out of the Patent Office and into one of the academic positions that had eluded him for so long.

Einstein received dispensation to take a break from work, and he and von Laue walked through the streets of Bern, going over the latest findings from Berlin, Heidelberg, and other important scientific centers. Einstein, as always, was puffing on a cheap cigar and was generous enough to offer one to von Laue. (Von Laue, used to better-quality tobacco, deftly managed to "lose" it over the side of a bridge.) But despite Einstein's overtures, and despite his polite follow-up letters, there still was no offer of a job after their meeting.

Einstein remained at the Patent Office, where, at an ordinary upright desk, he continued to labor for Superintendent Haller, as he had now for half a decade. In frustration he begged an old friend from his earlier days in Bern to move back and join him in the office. "Perhaps it would be possible to smuggle you in among the patent slaves," Einstein wrote enthusiastically. ". . . Keep in mind that besides the eight hours of work, each day also has eight hours for fooling around . . . I would love to have you here." His friend didn't take up the offer.

With his 1905 accomplishments receding and the Patent Office remaining a six-day-a-week job—and the only scientific library in Bern still closed on Sunday—Einstein could once again feel himself slipping away from the academic world. It wasn't that he hadn't tried to find another post. He knew that teaching in a high school would give him better hours, and in his agony at the Patent Office, Einstein peppered his friend Marcel Grossmann with questions about how to get a permanent job in a Swiss school. Would it matter that he spoke standard German rather than the Swiss dialect? Should he mention his scientific papers? Was he to call on the administrators in person, or would the fact that he looked Jewish get in the way? Whatever advice Grossmann gave him didn't help much. When Einstein did apply to a high school in nearby Zurich, his was one of twenty-one applications. Three applicants were selected for follow-up interviews. Patent clerk Einstein was not one of them.

Einstein also tried to teach at the University of Bern. On his first attempt, in June 1907, he was told that he needed to have finished a dissertation first. Since he didn't have a dissertation, he sent copies of his 1905 papers instead, at least three of which were worthy of the Nobel Prize. There was the September one laying out special relativity, as well as the November one showing how $E=mc^2$ was a consequence of it. But there was also a paper in which he produced a great understanding of photons. Possibly a fourth paper that year —building on simple microscopic observations to prove the existence of atoms—also deserved a Nobel Prize. But the university administrators wrote back explaining very clearly that perhaps Herr Einstein had not understood. This was Switzerland. There were bureaucratic requirements. He was obliged to send a dissertation, not some motley collection of papers. His application was rejected.

STUCK AT THE Patent Office, with only occasional visits from the likes of von Laue, Einstein didn't give up. He knew he was drawn to problems at the very limits of scientific understanding and that even the greatest minds made errors. He also knew that in 1905, he had already solved one of the great problems of science: why the universe was divided into so many separate "parts." His remarkable answer had been that it wasn't: that mass and energy are so deeply connected that each could be seen as a different aspect of the other. He had even revealed exactly how the universe arranged that interrelated mass and energy to shift back and forth. It was all there in $E=mc^2$.

By finding that such apparently unrelated items were interlinked, Einstein was primed for an encore that would take him to even greater heights. If all the mass and energy in the universe were interconnected—what we can informally think of as all "things" being interconnected—why did there still remain a seemingly separate domain of empty space? To have that second domain sitting there alongside those things of mass and energy—alongside all the universe's locomotives and planets and fire and stars—didn't seem very unified. Why should science screech to a halt without bringing all of space and all the "things" within it together as well, under the rubric of a single, grand theory?

Einstein began to wonder about the wider setting in which all the energy and all the mass—all the "things" of the universe—moved around. Something must be channeling them, guiding them. That appears impossible in the flat, empty space around us—but what if there was some explanation for how mass and energy moved around in this apparent void? What if space wasn't quite as empty, and flat, as it seemed?

To sensible thinkers, this quest seemed impossible. We know that the curve of an ocean swell can send a boat swerving to one side. But that makes sense because the swell is just the surface of a larger, three-dimensional body of water. That body is what the surface of the water is curving around. If Einstein's suspicions were right, however, and space is somehow curved, the question becomes:

What, possibly, is it curving around?

To understand Einstein's solution—and the confidence he gained from it, as well as the terrible errors to which it led him—it helps to turn to a quiet Victorian schoolmaster named Edwin Abbott. It was he who found that although it's impossible to visualize a dimension higher than the three we live amidst, it is possible to get a hint of how we might in fact be existing, unaware, within such a higher-dimensional universe.

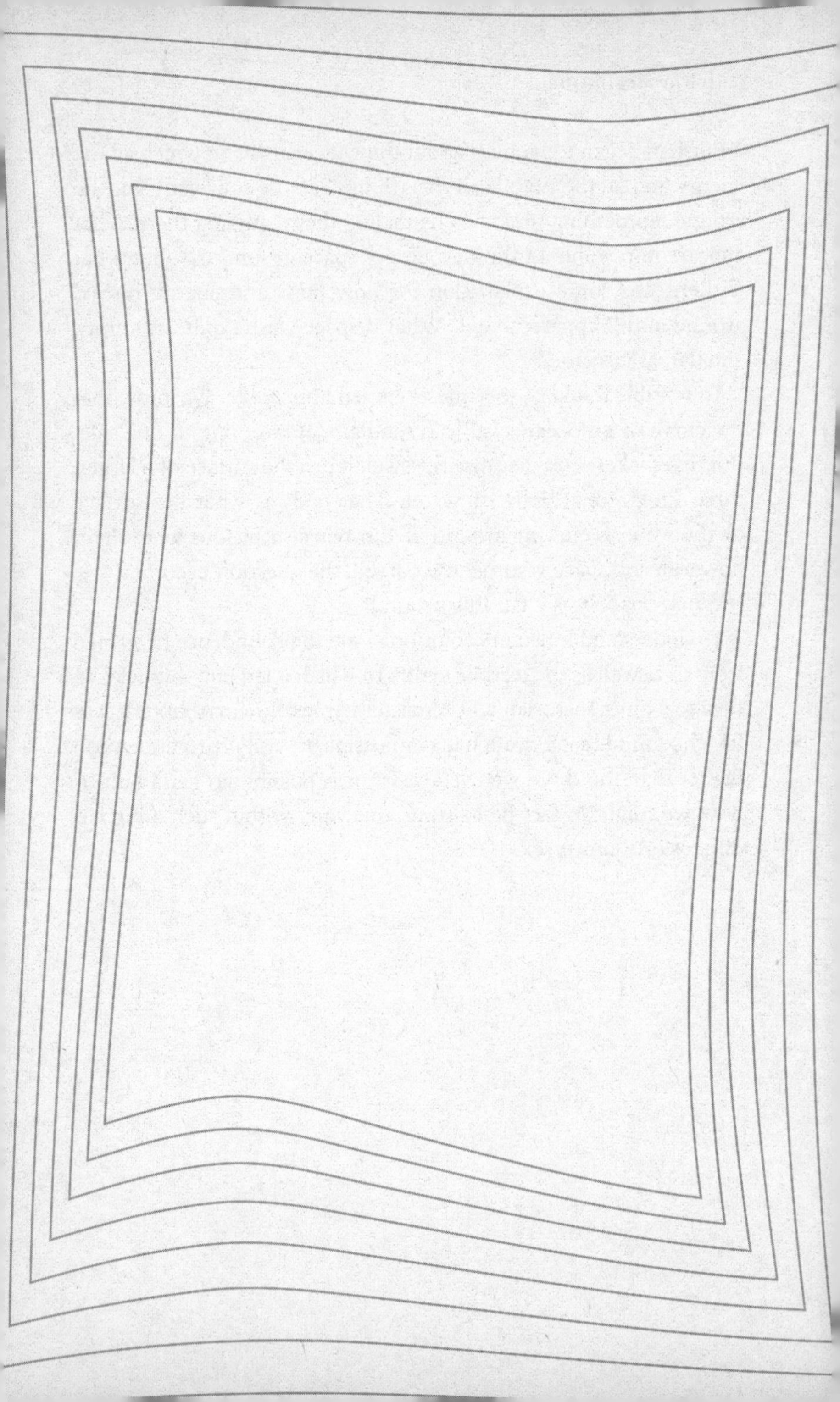

Part II

“THE HAPPIEST THOUGHT OF MY LIFE”

Einstein and Mileva Marić with their first son, Hans Albert, in Bern, around 1904

INTERLUDE 1

The Romance of Many Dimensions

IN 1884 EDWIN ABBOTT, then the head of the City of London School, did something that in Victorian society was more embarrassing for a distinguished schoolmaster than stepping out in the street without a hat. He published a novel that had a hero who was only eleven inches long and lived his entire life on a vast sheet of paper, "on which straight Lines, Triangles, Squares . . . Hexagons, and other figures move freely about, on or in the surface, but without the power of rising above or sinking below it." This world was called Flatland, and its author, as Londoners learned when the book was first published, was "A. Square," Abbott's pseudonym.

The book was a social satire, and it proposed an ingenious way to imagine a physical world that we can't see.

The lowliest of the beings living in Flatland are the straight lines, whose sharp, piercing tips have to be avoided at all costs. One social level above them are the workers: long, narrow triangles eleven inches on their main sides—beings of little education and dangerous if provoked, but usually docile enough to do what their betters tell them to. One level above *them* are the middle-class professionals—doctors, teachers, and other respectable fellows. They have the shape of squares, and the book's humble narrator is one of them. Another level up are the elite, who have yet more sides—pentagons, hexagons, and the like. At the very peak of society are the priestly circles, who glide wherever they wish along the surface, with lowly lines and pointy triangles taught to steer clear of them.

When the story begins, Mr. A. Square is fairly content with this flat world, though he is troubled by a dream he once had of some strange other world where all creatures lived on a single, one-dimensional line, existing like tiny trains restricted forever to a single track. Those poor beings could understand the idea of moving forward and back, but unlike Mr. A. Square, they could not conceive of there being an additional, "second" dimension that allowed movement from left to right. When Mr. A. Square crossed into their line, they could see just fragments of him as different points along his two-dimensional body entered and then departed their one-dimensional world.

A. Square's dream made it clear to him that visitors from higher dimensions possess a greater power than those in lower dimensions. If a being such as A. Square reached into the line he visited and plucked one of the creatures from its position, the locals left behind in "Lineland" would have no idea where their fellow had disappeared to. Then if A. Square put the Lineland creature back but in a different position, they would be mystified as to how he could appear in a new location without having traveled through the intervening space in any way they could tell.

When A. Square woke up from his dream and saw he was back in proper Flatland, he was content for a time. He was a prosperous enough man, with his own, impressively two-dimensional home: one with an opening for himself and his sons, as well as—for Flatland was a sexist society, and women were considered inferior—an extra, much smaller door that his wife and any other women were to glide in through.

All would have remained fine, but then, as Mr. A. Square remembered from prison later:

"It was the last day of our 1999th year of our era. The pattering of the rain [which strikes only at the wall of their houses, for there's no concept of such a thing as a roof] had long ago announced nightfall; and I was sitting in the company of my wife, musing on the events of the past and the prospects of . . . the coming century."

There was a strange sound in their house, and then suddenly, "[to] our horror . . . we saw before us a Figure!" It hadn't glided in through one of the two doors that led into the house. Rather, in some way that neither A. Square

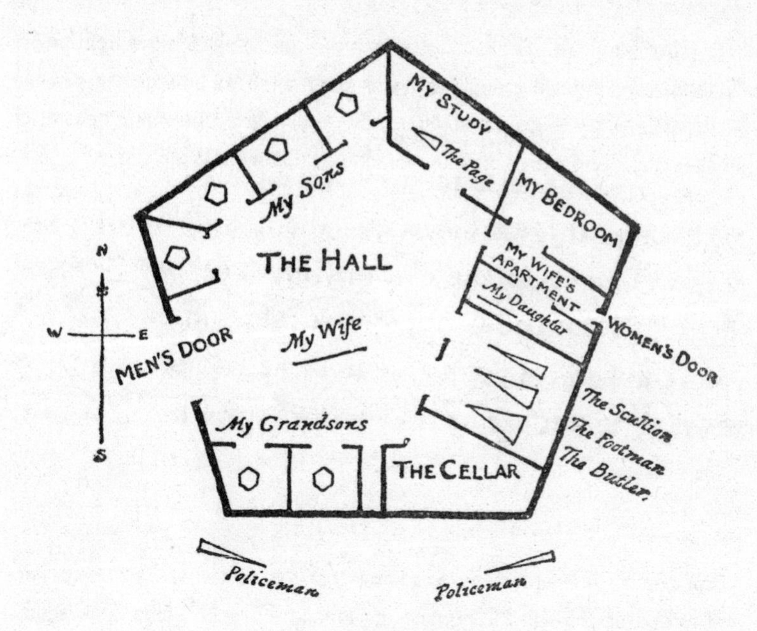

Mr. A. Square's house

nor his wife could fathom, it had just suddenly appeared in their room. The strange visitor quickly began to transform, from being a very small circle to a larger one. A. Square's wife was terrified, declared she had to go to bed, and glided as quickly as she could out of the room. Mr. Square was left alone with the stranger. With suitable politeness, he asked where his esteemed visitor had come from. The stranger said, "I came from [the Third Dimension]. It is up above and down below."

A. Square was confused. Surely, he told the visitor, he must mean that he came from north or south, or possibly from left or right. But the visitor was insistent: "I mean nothing of the kind. I mean a direction in which you cannot look."

A. Square thought this must be an attempt at a joke, but the visitor again was adamant: "Sir; listen to me. You are living on a Plane. What you style Flatland is the vast level surface . . . [on] the top of which you and your countrymen move about, without rising above or falling below it."

To prove his point, the visitor said that he was going to move from below Flatland, travel through it, and then hover over the top. As that happened, what A. Square saw astonished him. From being a big circle, the visitor became a smaller circle, and then a smaller one, ending up as just a tiny dot.

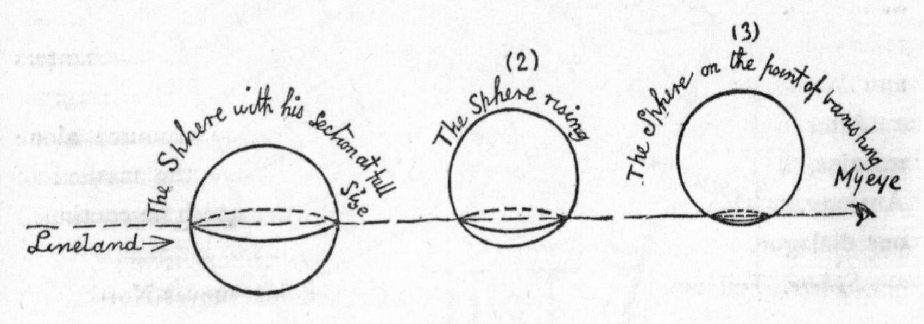

The sphere arriving

The process then reversed. We know that the visitor was a sphere who had moved up through the surface of Flatland and then come back again. Mr. A. Square, however, had only been able to see a series of sideways cuts. He was flummoxed. It was no surprise to him when the creatures in the one-dimensional Lineland were startled to see a fresh line suddenly appear in their midst. That was because they didn't understand that they actually existed on the wider, two-dimensional Flatland. But Mr. A. Square was convinced that was as far as matters went. He couldn't imagine that he himself actually existed within a broader three-dimensional space.

The visitor realized he needed to give a further demonstration. Mr. Square kept his account books in a large room (the study) in his house. The strange visitor asked Mr. Square to close and lock the door to that room. He then said he would rise up into a third dimension, which existed invisibly "above" Flatland. From there he would descend into the locked room (which had no roof, of course, for in a two-dimensional world, no such thing would exist) and take the account books.

Mr. Square didn't believe him. It was true that in his own dream about Lineland, he'd been able to reach in and grab things that to the lowly line creatures seemed suddenly to disappear. But that was because he could travel all around them in the exciting two-dimensional Flatland and they were stuck in a limited one-dimensional space. But nothing similar could happen here, he was sure, for what could possibly exist beyond Flatland? As the visitor got smaller and then disappeared, Mr. Square sprang into action.

"I rushed to the [study] and dashed the door open. One of the tablets was gone. With a mocking laugh, the Stranger appeared in the other corner of the room, and at the same time the tablet appeared upon the floor. I took it up. There could be no doubt—it was the missing tablet. I groaned with horror, doubting whether I was not out of my sense."

And at that moment, A. Square was finally ready to grasp the truth. The strange visitor explained.

"What you call Space is really nothing but a great Plane. *I* am in [true] Space, and look down upon the insides of the things of which you only see the outsides. You could leave the Plane yourself. A slight upward or downward motion would enable you to see all that I can see."

The stranger went ahead and lifted him "up."

"An unspeakable horror seized me," Mr. Square remembered. "There was a darkness; then a dizzy, sickening sensation."

The sphere told him to open his eyes and try to look steadily. And when he did that:

"I looked [down], and, behold, a new world! . . . My native city, with the interior of every house and every creature therein, lay open to my view in miniature."

He finally saw that the entire world he'd known before was composed solely of little geometric shapes, sliding around on the surface of a flat sheet. When he had been living down there, he hadn't recognized it, for this only made sense when he moved up to the higher dimension. It's a general principle: what seems odd to a creature living in a given dimension makes every bit of sense if that creature can envisage it from the next higher one. Creatures disappearing from Lineland and reappearing in new locations were baffling to the residents of that straight-line kingdom but made sense when viewed from Flatland. Similarly, what A. Square experienced with his visitor—objects magically disappearing from locked rooms—made sense once he realized he existed not just in the Flatland he was used to and could see, but that it was just part of a far greater Sphereland, which he had never been able to imagine.

Once back home, however, A. Square couldn't get his family or anyone else to grasp what he'd seen. As time went on, he also realized, to his distress, that he was starting to forget his eye-opening experience: "About eleven months after my return from Spaceland, I tried to see a Cube with my eye closed, but failed; and though I succeeded afterwards, I was not then quite certain (nor have I been ever afterwards) that I had exactly realized the original. This made me more melancholy than before."

A. Square's story did not end well. He was eventually brought before the High Council, where he found out that the priests of Flatland knew that they existed in only two dimensions. But since they didn't want to let the news get out—from fear of rebellion—and since in their eyes Mr. A. Square was not to be trusted, our intrepid explorer got locked up.

"Seven years have elapsed and I am still a prisoner," he says on the book's final page. His only hope is "that these memoirs, in some manner, I know not how, may find their way to the minds of humanity . . . and may stir up a race of rebels who shall refuse to be confined to limited Dimensionality."

THE ANALOGY OF *Flatland* is, of course, to our own world. Abbott wanted Englishmen to question the ruling class's privileges, which were so taken for granted that they often seemed entirely invisible. The straight-line fragments living in Lineland couldn't see that there was a wider two-dimensional world beyond them. The squares and pentagons and triangles living in Flatland couldn't see that there was a wider three-dimensional world beyond *them*.

This is why readers shouldn't feel bad for being unable to visualize curved space. No one can visualize it, not even an Einstein. Abbott simply wanted to say that even our greatest scientists might be as blinkered as the civilization of A. Square's Flatland. As Abbott was also a devout Christian, he didn't mind if readers saw parallels with religious beliefs—the arrival of the Logos in John 1:1, the miracles, the Ascension—that could seem impossible when limited to three-dimensional space.

It was around the time of the publication of *Flatland* that speculations about different geometries entered popular culture. In the Sherlock Holmes stories, the dastardly criminal Professor Moriarty is an expert in mathematics and would probably know of non-Euclidean geometries. In Dostoevsky's *Brothers Karamazov,* when Ivan tries to explain the problem of evil to his simpler brother Alyosha, he says, "I have a Euclidean earthly mind, how could I solve problems that are not of this world? And I advise you never to think about it either, my dear Alyosha, especially about God, whether He exists or not. All such questions are utterly inappropriate for a mind created with an idea of only three dimensions."

To most physicists, however, the issue of whether different geometries actually exist remained meaningless. Ivan Karamazov was a character from Dostoevsky's imagination. Professor Moriarty didn't exist. Scientists could get on with their work, unperturbed by the visions that had troubled the once contentedly bourgeois A. Square.

The hidden world that these beings had glimpsed, however, was exactly what Einstein would need to confront if he was going to solve the problems he'd begun struggling with at the Patent Office after he'd finally recovered from the exertions that had led to his $E=mc^2$.

FIVE

Glimpsing a Solution

By 1907 two years had passed since Einstein had published his series of papers—two years since he had united the realms of mass and energy, showing they could be seen as just a single category of interconnected "things," transforming when they need to in crisp accord with his equation $E=mc^2$.

Einstein's theories were powerful, to be sure, but they left open the question of why the unity in the universe didn't go further, and that question remained unresolved in 1907. All those "things" of mass and energy exist in a surrounding realm of "empty space." Why should God—or whatever forces set up the universe—have decided that there should be two utterly unrelated categories: "things" on one side, and "empty space" on the other? If energy and mass were interrelated, why wouldn't things and space be as well?

To Einstein, still rooted in a religion where one single deity created everything, it made no sense. So he got back to work.

The new project that Einstein began in the Patent Office in 1907 would yield a new theory. This one would be called general relativity, in contrast to the more restricted work he'd published in September and November 1905, which dealt with special relativity and its consequences. Einstein's second, broader effort would revolutionize physics in ways we are still grappling with today. This period of his life would lead him to creative heights that far surpassed his $E=mc^2$—but it would also, ultimately, lead to his fall.

GENIUS OPERATES IN indirect fashion. At work, Einstein liked to close his eyes, to tune out the scraping of fountain pens in his office and the constant tsk-tsking of Herr Haller as he patrolled the worktables, so that he could think more clearly. But at one point in 1907, he'd kept his eyes open as he was reflecting, and either had seen some workmen climbing on a ladder to the edge of a nearby roof, or just imagined them on the roof. In some unfathomable mix of neurons firing, he later recalled, suddenly "there came to me the happiest thought of my life."

Einstein began to think about falling from a house roof. If the house was very high, once you fell past the edge, neither you nor anyone else falling with you would be able to tell, without looking at the surroundings or feeling the wind, if you were moving or not. If you were holding hands with your partners and then let go, your partners would remain in the same position, as seemingly "stationary" as you were. You'd feel weightless, and so would they.

This would be your perspective as you fell. But if someone on the ground were looking up, not only would that person see you quickly plummeting downward, but he himself, of course, wouldn't be weightless. He would weigh just as much as he did before you slipped off the roof.

Why, Einstein wondered, should the person on the ground feel gravity and you suddenly *not* feel it? Gravity couldn't have suddenly disappeared around you when you slipped off the roof.

There had to be a way of understanding this better, and Abbott's book *Flatland* provides a start. Many of the characters in the book exist embedded in higher dimensions than they recognize, which means there are guiding "curves" in their own dimensions that explain what otherwise might seem mysterious. Consider the lowly beings who live in one-dimensional Lineland, existing like tiny trains on a narrow track. Their greatest geniuses would be perplexed if they found that after great travels, which constantly went straight ahead, they somehow ended back exactly where they had begun. But that would make perfect sense to an observer from a higher dimension, such as Mr. A. Square, who saw that the train track the Lineland beings lived on was actually curving in two-dimensional space and formed a circle. "We are," as Abbott put it in *Flatland*'s introduction, "all liable to the same errors, all alike the Slaves of our respective Dimensional prejudices."

The conclusion is straightforward. If objects move through higher dimensions, they can be guided in ways that seem incomprehensible to them. Here on earth, in our three-dimensional universe, we *think* that an invisible force of gravity is stretching up from the center of our planet and pulling us downward. But what if what's really going on is that when we fall, we're gliding along some curved pathway in space—a curve that's impossible for us to sense directly, but that mathematical analysis might be able to reveal? That would be a fantastic link between Things and Space: some sort of twist or channel existing in Space that Things slide along as they move.

The great Sir Isaac Newton was never convinced that he really understood how gravity worked. If Einstein could develop his own ideas about invisible channels in Space guiding our every movement, including our tumbling falls in gravity, he would have surpassed Newton.

This was a fabulous prospect. In a major review article in 1907, he began to expand his 1905 work on special relativity to include some of these new thoughts on gravity, but he had to stop before he'd properly developed his ideas. The Patent Office was proving to be an impossible place to work. This wasn't because he was overly delicate about needing silence to concentrate. Even among a noisy group, Einstein had the ability to, as Maja described it, "withdraw to the sofa, take pen and paper in hand, set the inkstand precariously on the armrest, and lose himself . . . in a problem." At one point when he was in his twenties, a visitor to Einstein's apartment described him sitting in a big chair, rocking his child in his left hand, writing his equations on a flat surface with his right hand, and keeping a cigar lit as he puffed away over infant, equations, and new visitor alike.

The links between Space and Things, however, were just too much to work out in those scattered evening hours available to him. Sometimes he could still evade Superintendent Haller and sneak open the drawer of his desk to take out papers from his self-styled Department of Theoretical Physics. But Haller seemed to be keeping a stricter eye on his clerks, and far too often Einstein had to slam his drawer shut before he'd had time to do any serious work.

There was also now another, more personal reason for seeking a better job. Although von Laue's visit in 1907 hadn't led to another position, as 1907 turned into 1908, Einstein's reputation was growing, and more visitors were starting to come. These weren't like the friends he and Marić had acquired together in their first years of marriage—friends they would stroll or share meals with. Nor were they like Grossmann, from the Polytechnic, with whom the couple could reminisce about their student days. The new visitors came to talk to Einstein, and to him alone.

Marić was no longer a fellow science student, far brighter and more educated than almost any other woman they knew. She was merely Mrs. Einstein, to be treated politely as she served beer or tea, but then ignored.

That was hard for Marić. She wasn't at Marcel Grossmann's level of mathematical dexterity, but she'd been a strong student, quite comfortable with advanced calculus, statistical mechanics, and the like. At that time, she and Einstein had dreamed of working together. Even as late as 1905, she had checked the most important of his articles, since he trusted her sharp eye for mathematical errors. When the last paper was finished, they had gone out and celebrated not like earnest parents, but like exuberant students again, as their card about ending up dead drunk showed.

Marić tried to fight her sadness at the change, writing to a girlfriend: "With that kind of fame, he does not have much time left for his wife . . . But what can you do?" It must also have hurt that a husband-and-wife team, the Curies, in Paris, *had* just won the Nobel Prize—exactly the dream that had had to be put aside when Marić needed so much time taking care of her and Einstein's son.

More money for child care would only help to free up both Einstein and Marić for the work they craved. So in the end, Einstein swallowed his pride and contacted the University of Bern again. They had rejected his first application for a teaching post, since his submission of the special theory of relativity didn't fit their requirements. Now he submitted the more conventional dissertation they wanted, and was accepted to give the very lowest level of lectures. There would be no pay, aside from what students who attended his lectures might contribute. He still had to continue at the Patent Office, but it was a start.

His first class was in the spring of 1908, meeting on Tuesday and Saturday at the desperately early hour of 7 a.m. When it looked as if no one would show up, the ever loyal Michele Besso, as well as two more friends from the Patent Office, decided to attend. Once the day's lecture was done, he would join them to grab a quick coffee, then hurry down the hill to work.

In the winter term the following year, they were joined by a real student, which was pretty exciting, but when that student quit it seems that Einstein's sister, Maja, promptly showed up to the lectures to keep the university authorities from canceling her brother's course. She understood not a word of what he said, and since Einstein wasn't going to charge Besso or his sister, money at home remained tight. "Isn't it clear to anyone that my husband works himself half dead?" Marić loyally replied when a friend commented that they should hire a maid so that she could have more free time.

Thankfully, before long news came that a properly paid position might be available at the University of Zurich, just sixty miles away. That, however, required having a professor from Zurich come and watch Einstein lecture. This was a worry. For Einstein it was always hit-or-miss whether he would give a good lecture, "due to my poor memory." When the great day arrived and Einstein got home afterward, Marić asked how it had gone. The news wasn't good. "Being investigated got on my nerves," he explained. "I really did not lecture divinely."

Eventually Zurich relented, largely because an ever greater number of physicists across Europe were recognizing the strength of Einstein's papers. Moreover, when the in-house physics candidate realized that the faculty might still reject Einstein, that candidate—Friedrich Adler, an old acquaintance from Polytechnic days—to his credit stepped back: "If it is possible to get a man like Einstein for our university, it would be absurd to appoint me."

So it was in 1909 that, after a biblical seven years of servitude, Einstein finally was able to leave the kingdom of the patent slaves and take up his first proper academic job at a university. Haller seems to have been almost entirely oblivious to Einstein's growing fame, and in line with standard bureaucratic work had merely promoted him to the position of Technical Expert Second Class, although before Einstein left—perhaps in an effort to keep him on?—he hinted that the exalted heights of Technical Expert First Class might someday be within reach. By leaving Haller's office, however, Einstein would now, at last, be able to continue his investigation: to see if the universe's deepest parts really were linked, by curves or paths that no one had imagined before.

SIX

Time to Think

IN 1909, THE YEAR he moved from the Patent Office to the University of Zurich, Einstein was thirty; Marić was thirty-four. Bern had been attractive, but it was also isolating: it was really just an overgrown small town. Zurich was a true city, and many of their friends from Polytechnic days still lived there. That fact alone seemed to bode well.

The move proved rejuvenating, and for a while life was as thrilling as when they had first been married. They met Carl Jung, which could have been excellent for Marić, as her first interest, before physics, had been medicine, and so they potentially had a lot in common. But when Jung invited the Einsteins over for dinner, he largely ignored Marić and just focused on Einstein, trying to convince him of his own psychological ideas. Einstein didn't enjoy that, and they never went back.

The Einsteins had better luck with a specialist in forensic medicine at the university, Heinrich Zangger, an ingenious man—one of the founders of emergency room medicine—whose range of interests greatly impressed Einstein. Better still, the Einsteins moved into the same block as the family of Albert's academic advocate Friedrich Adler, who noted the good mood in the couple's apartment.

"We are on very good terms with [the Einsteins], who live above us," Friedrich wrote his father. ". . . They run a Bohemian household."

The University of Zurich salary was better than that at the Patent Office, but Einstein and Marić both knew it was important he didn't get fired for lecturing poorly. He still dressed differently from the other faculty members in Zurich, with his trousers too short and his hair rumpled, but both he and Marić liked the idea that they were far from an ordinary, bourgeois couple. Einstein prepared his lectures more thoroughly than in Bern, and instead of trusting his poor memory, one student remembered that Dr. Einstein brought "a scrap of paper the size of a visiting card, on which he outlined the ground he intended to cover with us."

Most of all, Einstein treated his students kindly. Europe before the First World War had strict hierarchies, and professors didn't invite questions, certainly not from ordinary students. Einstein, however, had always scorned people who put on airs because their social status allowed it. Here in Zurich, he encouraged his students to interrupt him whenever they had a question; he invited them to coffeehouses after class to continue the conversations, or just to get to know them; often he'd bring them home to share his latest research. They liked it. He also always stood up against bullying. One student, a few years later, remembered how nervous she had been before delivering a seminar. Einstein gave her a reassuring nod from the audience, as if to say "Go ahead, you'll be fine." When a brash male student tried to score points by putting her down, Einstein stopped him, saying, "Clever, but not true," and then encouraged her to continue.

The Einstein family's new Zurich apartment was bigger than the one in Bern, and with that space, plus their renewed affection, they soon had a second child, named Eduard. One visiting student remembered that when the two boys made too much noise for Einstein to concentrate, the young professor would smile, reach for his violin—this good dad's surefire weapon—and put them at ease by playing their favorite tunes. He and Marić called their sons *die Barchën,* "the little bears."

In 1911 a better job came up, at the German University in Prague, and so the family moved again. Here Einstein's salary stretched to a truly grand apartment—their first with electric lights—and he had even more time, between his administrative duties, simply to think.

Prague was in some ways a respite for Einstein, but it was far less pleasant for the German-speaking Slavic Marić, not least because of the standoff between German speakers and Czech speakers in the city. Czech nationalism was growing, but the German minority had control of many senior positions. Czechs who were perfectly bilingual often refused to speak German, to discomfit those, like Marić, who dared to try shopping in their city without knowing Czech; German speakers, even more ominously, took to disparaging all Slavs, which of course included Marić. The very fact that there was a "German University" demonstrated the problem, for it had been created when a separate "Czech University" split off from the same faculty, and now—although Einstein made a point of opening his lecture to Czech students—most professors refused to speak to anyone at the opposing university. There was a small Jewish literary crowd in the city that tried to stay neutral. It was at one of their salons that Einstein met Franz Kafka, although it seems that Kafka was too shy to say anything to this easygoing, already respected foreigner. What they might have talked about otherwise, we can only imagine.

PRAGUE MAY NOT have been the easiest place for the Einsteins to live, but at least there Albert was able to take his thought experiments further. He already had some idea that space itself was distorted in some way, which would explain what he was imagining about gravity, but he couldn't yet work out the details. He also had some notion that because of these distortions, distant starlight would curve if it traveled near the sun, but he couldn't be entirely sure about the details of that either.

One thing that helped, curiously enough, came from a genre of adventure stories in which a heroic explorer gets drugged, then wakes up bewildered and has little time to work out where he is. Einstein used that idea. Suppose, he imagined, that someone did wake up in a closed room with no windows: that he had been drugged and has no memory of how he got there. He can't feel any gravity; he's just floating in the room.

Is there any way that he can discover where he is?

One possibility, the heroic explorer would realize, is that he's somewhere in distant space, out beyond the solar system and away from any large, bulky source of gravity such as our sun or even Jupiter. But another possibility is that he might just be in a building's elevator, like those in the new skyscrapers then being built in America, and a dastardly villain has cut the cable, so he's dropping from the very top of the elevator shaft. If the room is totally enclosed, and he's floating freely, he can't tell which it is. This is like the workers whom Einstein had imagined falling from the rooftop in Bern. When they're in the air, unable to look around or feel the air's movement, all they know is that they're weightless. Whether they're miles up or just inches from the ground, they cannot tell.

There is a way, however, Einstein now realized, that our intrepid hero can work out where he is, without being able to peer outside his sealed room. All he needs are two apples. He places one apple in each hand, spreads his arms out, and then lets the apples go.

If the two apples stay hovering perfectly still, he will know that he really is far away, in the distant vastness of outer space, far from any rock-jagged planets. He will have plenty of time to build an engine and get himself to safety.

But if the hero lets go of the apples, and instead of hovering in position, they ever so slowly, but ever so definitely, start gliding toward him—if he knows it's not an air current doing that, or his own tense exhalations—then he will realize he's in serious trouble. There's only one thing that can make two apples that begin quite parallel with him eerily begin to approach him. There has to be a central source of gravity somewhere below, one that each apple is aiming for from its own starting point:

One can see the effect more strongly if one imagines him above the earth:

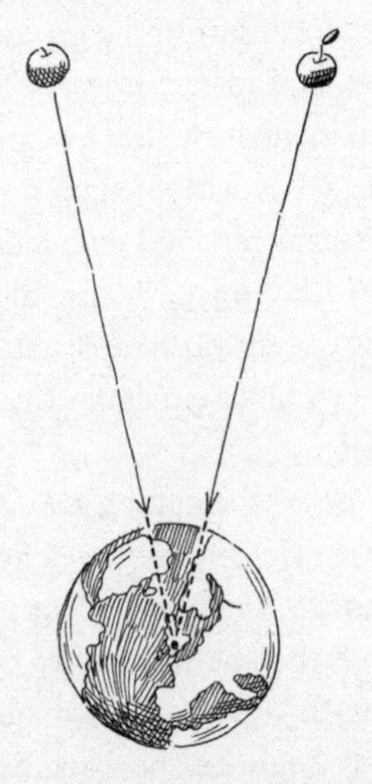

The conclusion is unfortunate, but also unambiguous. When the same effect happens in miniature, it's clear that our hero must be in the free-falling elevator. At any moment now, he, the apples, and the entire room are going to most painfully impact upon the ground.

Looking at how the apples move is an ingenious way for the explorer to deduce if he's moving toward a gravitational source like our planet, or if he's far away in distant space. But there's a conundrum. He feels no force whatsoever while he's free-floating. Yet something is making the equally free-floating apples move toward him, and if he feels no force, it's natural to think that they feel no force either.

How can the empty space inside an imagined elevator lead objects such as free-floating apples to start gliding toward each other, even though, to the explorer there with them, they obviously are just suspended in empty air?

In grappling with this problem, Einstein was learning a lot about his own creative process. Thinkers have often been classified as either golfers or tennis players. The golfers work on their own; the tennis players need interplay. Newton was a golfer; Watson and Crick—like many composers and lyricists—were tennis players. Einstein had been a golfer long enough. He could make some progress on this problem on his own, but he needed to collaborate if he was going to get any further.

Who could Einstein turn to? Marić couldn't help, for although she had been able to check his earlier papers, having reached a good understanding of mathematics and physics in her undergraduate studies at the Zurich Polytechnic, this problem went far beyond what either of them had learned there. Besso would be out for the same reason, for although he had been, as Einstein put it, "the best sounding board in Europe," his lack of ambition and whimsical attitude toward serious study meant that he, too, didn't know enough, or wouldn't learn enough, to be helpful.

The person Einstein really needed on the long, slow path that would lead him to general relativity was Marcel Grossmann, his friend and the fellow student who'd offered him lecture notes back when they were both undergraduates. After a stint as a high school teacher, Grossmann had gone to graduate school to study advanced mathematics, and since then he had stayed in academia, ending up as a math professor at their old Zurich Polytechnic, which had recently been upgraded to a full university and was now known by its German initials as ETH. The two men had been in touch a few times in the decade since, as when Grossmann helped Einstein get a job at the Bern Patent Office or assisted with his friend's abortive high school teaching applications, but mostly they had slipped apart. Einstein, however, still had the greatest respect for his talents. If a position could be had back in Switzerland, Einstein could benefit from being close to him.

The Einsteins had an additional and personal reason for planning a move. Leaving their Zurich friends behind had put too much pressure on their marriage. The coolness they had experienced in Prague from both Czechs and German speakers hadn't helped either, and both Einstein and Marić were feeling the distance. When a big conference had come up in Brussels, one bringing most of Europe's top physicists together, Einstein hadn't brought Marić, even though she could have been in the company of the great minds there she so admired: Ernest Rutherford from Manchester, Max Planck from Berlin, and, of course, the successful female physicist she had never had the chance to try to become, Marie Curie from Paris. Marić wrote her husband while he was away, the letter carried by the fast steam trains that crossed the continent: "I would have loved only too much to have listened a little, and to have seen all those fine people. It has been an eternity since we have seen each other . . . Will you still recognize me?"

Perhaps returning to Zurich would bring back the warmth they'd had before. Marić was only too happy, accordingly, when Einstein arranged a faculty post for himself at ETH, the institution that just a short time ago had wanted no part of him. The family packed and in 1912 moved back.

Shortly after they reached Zurich, Einstein burst into his friend's study and said, "*Grossmann, Du musst mir helfen, sonst werd' ich verrückt!* (Grossmann, you've got to help me, or I'll go crazy!)" Grossmann was willing to oblige. Einstein now had a convenient position at ETH, the old Polytechnic, right next to his old friend and supporter—and, now, fellow teacher as well.

SEVEN

Sharpening the Tools

GROSSMANN'S FIRST STEP was to help Einstein catch up with the mathematics he'd missed in their student days by skipping so many classes. If empty space had curves in it, they would need some way of detecting this. Einstein was astounded when Grossmann—this man seemingly knew everything!—showed him that many of the necessary tools had already been worked out.

The mathematical approaches Grossmann showed Einstein built on what cartographers trekking over our planet, measuring latitudes and longitudes, had long before realized. When eighteenth-century surveyors measured points from wooden observational towers that were dozens of miles apart, even if the ground in between seemed a flat, snow-driven waste, they were able to tell from the size of the angles how curved or not the surface really was.

On flat plain, any huge rectangle that was staked out would have all of its interior angles a strict, clean 90 degrees. On much more curved surfaces, rectangles are "pushed" upward in the middle, so the angles at their corners are swollen to more than 90 degrees.

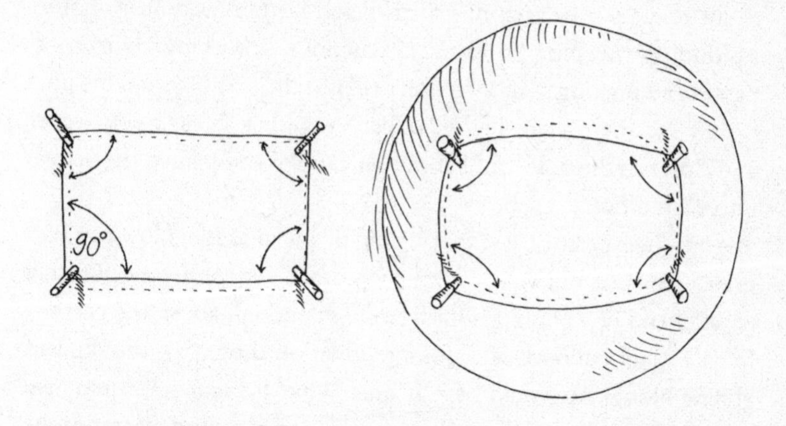

The earth's surface is a continuously curved surface, and while this is so gradual that travelers can't detect it with the naked eye, the curvature can produce amazing effects. Imagine, for example, that there is perfectly flat ice from Finland to the North Pole. Two skaters from a small town in Finland are made to stand a mile or two away from each other and then, when a signal is given, are told to skate on absolutely straight lines due north.

At first they think this is going to be easy. From their experience on the flat frozen lakes near their homes, they know that two skaters who start out in parallel can seemingly continue that way indefinitely.

But now, as they continue farther and farther away from home, assiduously keeping to their instructions—checking their compasses and making sure they don't veer to the side by even one inch—they see themselves being "pulled" together as each gets closer and closer to the pole, until at the very top of our planet, they crash into each other.

From their perspective, that would be inexplicable. How can two men who start out miles from each other and very carefully stay parallel end up colliding? But from high enough above the earth—from a giant balloon, say, looking down on those two tiny figures gliding along—it would be obvious. What the skaters experience as an ineluctable pull toward each other is not due to a mysterious force. Because the basic shape of the earth is a sphere, any travelers following straight, parallel lines on the curved surface will necessarily collide.

This phenomenon is identical to the one Einstein imagined in his thought experiment with the eerily approaching apples—it's just that one doesn't appear to take place on a surface, while the other one does. In Einstein's day, hardly anyone believed that these strange effects and curved paths could apply *outside* the surface of our planet—that outer space, which seemed empty, might also have some hidden structure that could influence how the objects within it moved. Rather, everyone assumed that the distant space in which planets and stars existed was as Newton had imagined it: flat, empty—a bare, dark stage set, before the actors appear.

Grossmann now explained to Einstein that a few mathematicians had already dared to look beyond that widely shared assumption. Several decades before Abbott wrote his *Flatland* fable, these intrepid few had begun to imagine that our planet might exist within wider geometries than we could possibly see. To the Hungarian army officer János Bolyai, writing in 1820, the idea was so exciting that after working through the logical possibilities, he wrote, "created a new universe from nothing!" To the academic German mathematician Carl Friedrich Gauss, who explored these ideas on and off for decades, "the theorems of [curved geometries] appear to be paradoxical, and to the uninitiated absurd; but calm, steady reflection reveals that they contain nothing at all impossible."

But when none of these elite mathematicians found the experimental evidence to back up these possibilities, the field dwindled. Abbott did pick up some knowledge of these abortive efforts when he studied at Cambridge, and there were occasional mentions in literature, but most physicists couldn't take this seriously. Mathematicians who continued to play with these possibilities were generally thought to be wasting their time. Even Einstein had joined in the mockery, writing to Marić back in 1902, "Grossmann is getting his doctorate on a topic that is connected with fiddling around with non-Euclidean geometry. I don't know what it is." Now, though, in 1912, Einstein changed his view. "I have become imbued with a great respect for mathematics!" he admitted.

THE LONG-FORGOTTEN TOOLS the pioneering mathematicians of the previous century had developed for studying these geometries of curved space were indeed tremendous—and they were also perfectly suited for the task Einstein and Grossmann had at hand. This was especially clear in an idea that one of Gauss's protégés, the mathematician Bernhard Riemann, had demonstrated in an 1854 lecture—one the elderly Gauss attended—in which he noted that creatures who lived on any sort of surface would be able to work out how much it was curved at any particular location. This idea developed what the cartographers had already noted: If triangles bulged outward, whatever surface they existed on was like that of our spherical earth. If triangles shrank inward, the surface was concave—*and all that could be seen without stepping off the surface.* Mr. A. Square, living in a two-dimensional universe, could have used these procedures to deduce that he was living on a flat surface even before the visiting sphere lifted him up to let him see it from above.

If we followed Gauss and Riemann's procedures carefully enough, we, too, Einstein realized—by measuring angles across great distances—could tell if something was making our three-dimensional space bulge or shrink. No one could detect it without such measuring equipment, since to our unaided senses the space right in front of us obviously seems flat. Humans have no capacity to "see" higher dimensions; not even Einstein. But with our calculations, we would be able to tell if there were "curves" there.

The underlying idea was so simple, and so beautiful, that Einstein later felt comfortable explaining it to his younger son, Eduard. Imagine, he said, that a small caterpillar is crawling around a large tree trunk. The caterpillar can't tell that the tree trunk under him is curved and that he's taking a curving path through space as he crawls. Only we, looking at the trunk from farther away, can see that taking place. The reason that Einstein spent so much time in his study, he explained to his son, was that he was trying to find a way for the caterpillar, caught in those paths, to work out if the world he was on was actually curved.

Einstein was still playing intellectual "golf" on the side, but Grossmann helped a lot as his occasional tennis partner. "I am now working exclusively on the gravitation problem," Einstein wrote the once suspicious but now admiring physicist Arnold Sommerfeld in Munich, "and I believe that, with the help of a mathematical friend here, I will overcome all difficulties."

Grossmann and Einstein made a good pair, even though they enjoyed playing up their differences. Grossmann was "not the kind of vagabond and eccentric I was," Einstein noted later. In their nearly two years together at ETH, Einstein lived in rumpled, comfortable clothes; Grossmann always wore a proper suit and a crisp, high-collared white shirt. Whereas Einstein teased that he'd stayed away from mathematics because "[it] was split into numerous specialties, each of which could easily absorb our short lifetime," Grossmann suggested that physics was ridiculously simple, saying that there was only one useful insight it had been able to teach him. Before studying physics, Grossmann said, "when I sat on a chair, and felt the trace of heat left by [the person before me] I used to shudder a little. That is completely gone. For on this point physics has taught me that heat is something completely impersonal."

Einstein's notebook from this period survives—a small, brown, cloth-covered volume filled with his neat inked handwriting, all the letters slightly angled to the right. On the first page, he doodles with recreational puzzles, drawing a system of train tracks and shunted railcars to help work through them. But then he gets into his serious calculations. After several pages, the plaintive words *"zu umstaendlich"*—"too complicated"—appear when Einstein finds himself stuck, trying to list curvatures in ways that would make sense from whatever direction an observer approached a surface. At another point, the name "Grossmann" reassuringly appears—just at the place where his friend has brought in a key idea to help.

In 1913 Einstein and Grossmann presented their preliminary findings in a paper with an appropriate two-part structure: Grossmann signed the mathematical part, and Einstein the physical part. But

Einstein's skill was improving. By the end of that year, he had arranged to take a full-time post in Berlin the next year. Grossmann had given him as much help as he could.

From here on, Einstein was on his own.

COMPLETING ALONE WHAT he and Grossmann had begun together was the hardest work of Einstein's life. "Compared with this problem, [1905's] original theory of relativity is child's play," Einstein wrote. "No one who had not gone through the torments, false hopes, could know what it entailed."

His colleagues saw how immersed he was. "Einstein is stuck so deep into gravity that he is deaf to anything else," Arnold Sommerfeld reported to a colleague. But Einstein kept at it as the months rolled on—"Never in my life have I tormented myself like this," he observed—because he felt something far greater than $E=mc^2$ was waiting to be discovered. "Nature is only showing us the tail of the lion," he wrote to his old forensic medicine friend Heinrich Zangger in Zurich. "But I have no doubt that the lion belongs to it, even though, because of its colossal size, it cannot directly reveal itself to the beholder."

There was a further complication. The move back to Zurich in 1912 had done nothing for his and Marić's marriage. Partly it was the sexism of the era, pushing the educated, intelligent Marić into a life focused on the home. Also, fatally, though still living with Marić in Zurich, Einstein had become entranced by a distant relative in Berlin, Elsa Lowenthal, a widow who had two grown daughters.

A trained actress with beautiful blue eyes, Lowenthal was well connected in Berlin's art world. She spoke French fluently, far better than Einstein (which wasn't especially hard; one sympathetic Frenchman he met reported that not only did he mangle the language by enunciating far too ponderously, but he also often mixed in some German). Lowenthal shared Einstein's appreciation of music and theater, yet she also knew him well enough to be amused when he mocked her more pompous friends. And having been educated in the arts, not the sciences, she had no reason to feel diminished if scientific visitors only briefly acknowledged her before turning to Einstein.

At one point in 1912, Einstein realized he had to stop all contact with Lowenthal, and wrote her a letter telling her so: his wife had begun to understand that she was not just a distant relative but a threat. But Einstein also included his return address, and when, early in 1913, she casually wrote to him on the pretense of wanting advice on finding a good popular guide to relativity, he couldn't resist starting their correspondence again.

Marić was furious when Einstein accepted the offer to move from Zurich to Berlin, for she knew it meant her husband would be closer to this woman who threatened their family. Their young sons had no idea what was going on, and when the family did transport all their belongings again, arriving in Berlin in the spring of 1914, they seemed delighted by the huge, modern city. But for Einstein and Marić, the days of delighting in new moves, of sitting on their balcony, looking at the Alps, and holding each other, were now impossibly far away. Friends saw how suspicious, cold, and easily hurt they were. In those first Berlin weeks in 1914, Einstein cruelly told Marić that he would put up a minimal front of friendliness only if it was "completely necessary for social reasons," even though the impending breakup was clearly his fault.

By July 1914, it was too much. Marić couldn't live like this, with her husband so clearly enamored of someone else. She still thought their marriage could be saved, but she had too much pride to stay. Einstein was caught, for in his heart their marriage was over—he'd even begun to refer to Lowenthal's children as his stepdaughters—yet he also wanted to keep seeing his sons. In the end, the warm-hearted Besso traveled from Zurich to help Marić and the boys move back to Switzerland. Einstein didn't insist on a divorce and agreed that he would send her half his salary. Sobbing at the Berlin train station as he watched his children go, Einstein then found a small apartment for himself with just enough space for the boys to visit.

The breakup was exhausting, as was his continuing work, and that was not all: the month after he and Marić had split, war had broken out in Europe. Conditions in Berlin rapidly deteriorated. Soon food was limited, there were cuts in electricity and fuel, and a frantic nationalism took over. To his old friend Besso he wrote, "When I talk to people I can sense the pathological in their state of mind." To a friend in the Netherlands he elaborated, "I am convinced that this is some kind of mental epidemic."

Einstein's life was in chaos, but how could he let his explorations go? He had to solve the problem of gravitation that he had been struggling with on and off since 1907; he had to unveil the universe's innermost secret.

And then, in November 1915, he cracked it.

EIGHT

The Greatest Idea

WHAT EINSTEIN DISCOVERED, in the chill of wartime Berlin, was the greatest breakthrough in understanding the physical universe since Newton: an achievement for all time. If Einstein had never been born, almost certainly someone else would have come up with $E=mc^2$, and not much later than he did, in 1905. The Frenchman Henri Poincaré and the Dutchman Hendrik Lorentz, for example, were at most a few years behind him. But no one else had come close to what Einstein achieved in 1915. Although the details are subtle,* the core can be represented as follows.

Think of truly empty space as like a vast trampoline surface. It's flat; there's no curvature, no dips or rises. If you flick a tiny ball bearing along the trampoline surface, it doesn't distort the trampoline at all, and just travels in a straight line.

Now place a small rock on that surface. Its weight makes the trampoline sag downward. Flick the ball bearing again, and if it passes anywhere near that rock, it will veer slightly toward it be-

* The appendix goes into this in more detail. In particular, we'll see that it's not just space that gets curved, but time as well.

cause of the sag. The mass of the rock makes the trampoline distort, and that distortion shifts the path of other objects—such as the ball bearing—that come anywhere near, as shown:

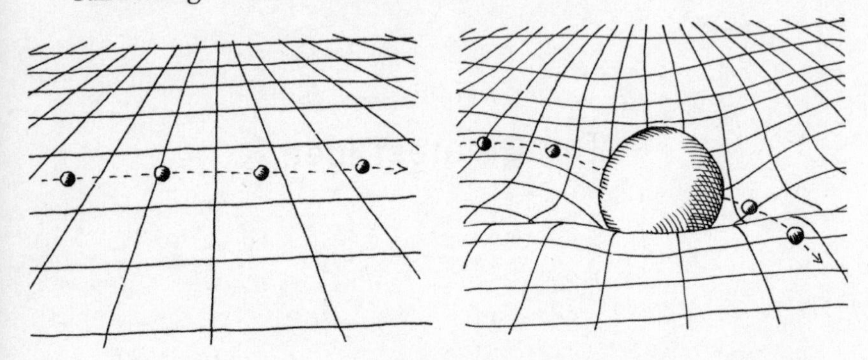

That was Einstein's vision; that was the theory that explains where the distortion of space is coming from. The warping—the curvature he had struggled to define since thinking about the adventurer in the elevator—comes from all the things, the stuff—all the mass and energy—that are scattered through space! Wherever mass or energy is located, they distort the space around them, just like the rock pushing down on our trampoline. Put a small mass somewhere new—move the mass of your hand a few inches sideways through the air, for instance—and it's as if you're pressing on invisible rubber sheets, and there really now is a slightly different configuration of space around it. Have a large mass arrive somewhere new—have the entire earth rush forward in its orbit—and it'll produce far bigger distortions in the invisible space around us.

It was a brilliant, bold idea, and in many ways a parallel to Einstein's earlier work on locating the tunnel between the two domed cities of *M* and *E*. Just as energy and mass were connected through an invisible linkage, Einstein realized, so were those two things interwoven with the space they occupied. There was a unity to the universe, he had always believed, and now he was one step closer to describing it.

Einstein's theory about the distortion of space represented a watershed in the history of physics, yet it was only half of what he had discovered. For in pinpointing the effect that things had on the space around them, he'd also gained new insight into how that influenced other things in their vicinity.

After all, what happens when a trampoline is bent and sagging? Those distortions in its geometry make the objects near them veer and shift. A ball bearing on the sagging trampoline isn't tugged by any mysterious force from the rock. It's simply following the most straightforward path from its perspective.

The idea makes intuitive sense. Create a distorted geometry at some point, and that will lead any things that are nearby to follow a new, otherwise inexplicable path. As we saw, that is why the two Finnish skaters would find themselves ineluctably drawn together as they approach the North Pole. They're skidding on a two-dimensional surface, which curves around our three-dimensional planet. That is also why the two apples let loose in the free-floating room slowly start moving toward each other if there's a source of gravity below. They're skidding along on a three-dimensional space, which —by Einstein's idea—must be the curved surface of an invisible-to-them four-dimensional space that it wraps around. The unhappy explorer floating between them is simply seeing them tumble along that curve.

In Einstein's radical reconception of space, there's no need to imagine an additional force of gravity; rather, gravity is simply the result of the warping of space. The snowy North Pole is not sending out an invisible force tugging the skaters closer to each other. Unless something pushes them away, objects *always* follow the most straightforward channels stretching ahead of them. One doesn't even have to envisage the icy north or falling rooms. Watch a surfer rise several feet up on the sea. If the ocean swell under him were invisible, this rise upward would be a great mystery, as would the subsequent glide downward. The moment you see that water, however, all is evident.

Both the geometry of space and the movement of objects within it, Einstein realized, are determined by distortions in space caused by the objects themselves. If space has nothing in it, there's no distortion at all: it's like a flat, geometrical plane. If there is a single planet on that plane, then there will be some distortion, as the planet makes the space around it sag downward. There will be even more dips and distortions if there are dozens of planets, all tugging the space around them.

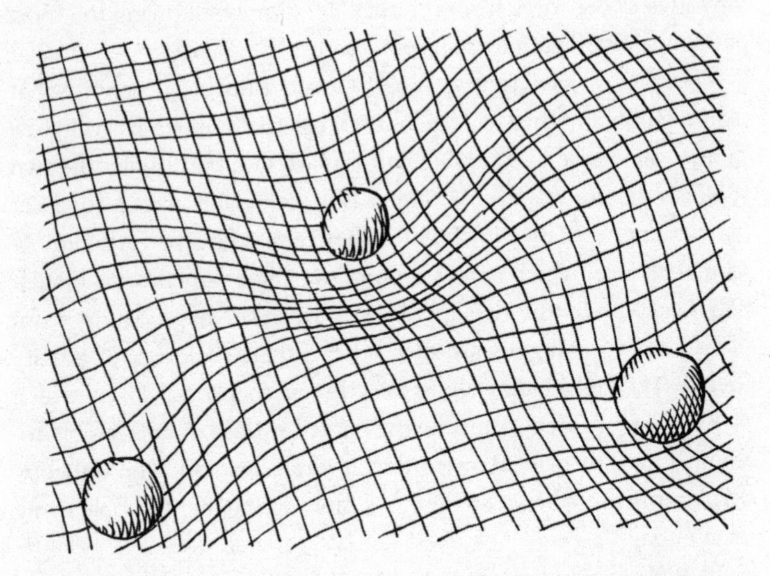

This realization fundamentally changed the way we understand the fabric of our universe. In 1816 the mathematician Gauss had written, "Perhaps in another life we will be able to obtain insight into the nature of space, which is now unattainable." Less than a century later, Einstein had done that. The domain of Things, and the Geometry of the domain of the space surrounding it, were not separate after all. There was a deep link. Place a huge agglomeration of rock into a new spot—let the vast bulk of the earth itself take up some position in our solar system—and its tremendous mass sags space enough to press humans, and apple trees, and entire mountain chains tightly to the surface; enough to guide aircraft, and space shuttles, and even the distant moon as well. That's how our sun leads the earth along: it's as if the sun has opened up a furrow around it, and we whirl along within it. The reason we feel at every moment as if we're going straight forward is that we're unable to step way back and "see" that giant curving path that we're gliding along. Yet, in his mind, Einstein had done just that.

SYMBOLS ARE MORE precise than words. To say "mass and energy lead to a sagging of space" is only a very rough approximation. What Einstein wrote can, still roughly, be better expressed as saying that there are "things" at one location—what we can call *T* for short. And there's a distorted geometry that those things produce near them. Let's call that distorted geometry G for short.

Einstein's insight—as summarized in the trampoline drawings—is that any arrangement of things (*T*) produces a distinctive new geometry (*G*) around it. The groups of things that exist at some location—be they hands, or mountains, or exploding flares—make the geometry around them bend or shift. A change in *T*, in other words, leads to a change in G around it.

The simplicity of Einstein's realization was stunning, and it came out in an extremely "brief" equation. How to tell how things are going to move? Simply look at the distorted geometry of space around them. In successively briefer phrasing:

- The geometry of space—our sagging trampoline—guides how things move.
- Geometry guides Things.
- G guides *T*.
- $G \rightarrow T$.
- $G = T$.

And how to tell *how* space is contorted? Simply look at the things sitting in it. Again, in briefer and briefer phrasing:

- Things distort the geometry of space—of our sagging trampoline—around them.
- Things distort Geometry.
- *T* distorts *G*.
- $T \rightarrow G$.
- $T = G$.

How extraordinarily symmetrical the equation at the heart of how our universe configures itself turns out to be! Almost the whole structure and dynamics of the universe are there, in just two neatly balancing phrases. Things distort geometry. Geometry guides things. Use the equal sign as a shorthand to summarize those mutual operations, and briefest of all, combining both, one gets a single equation: G=T. Einstein's symbols had further detailed characteristics, but although G=T is only a metaphor, it's a quite close one, and matches the essence of what Einstein wrote.

It was a fabulous discovery: that what seems odd and random to our senses, such as the tumbling of planets through space, actually comes from very clear, very exact laws. Best of all, human reasoning was able to uncover it.

Einstein tried to be modest about this equation, which would become the core of his general theory of relativity. He later said, "When a man after long years of searching chances on a thought which discloses something of the beauty of this mysterious universe, he should not be personally celebrated." But at the time, he couldn't resist. In 1915 he wrote exuberantly, "[This] is the greatest satisfaction of my life." And to his friend Michele Besso, he was even less guarded. "My boldest dreams have now come true," Einstein wrote after he had cracked it in November 1915, before signing off, "regards from your content, but *kaput,* Albert."

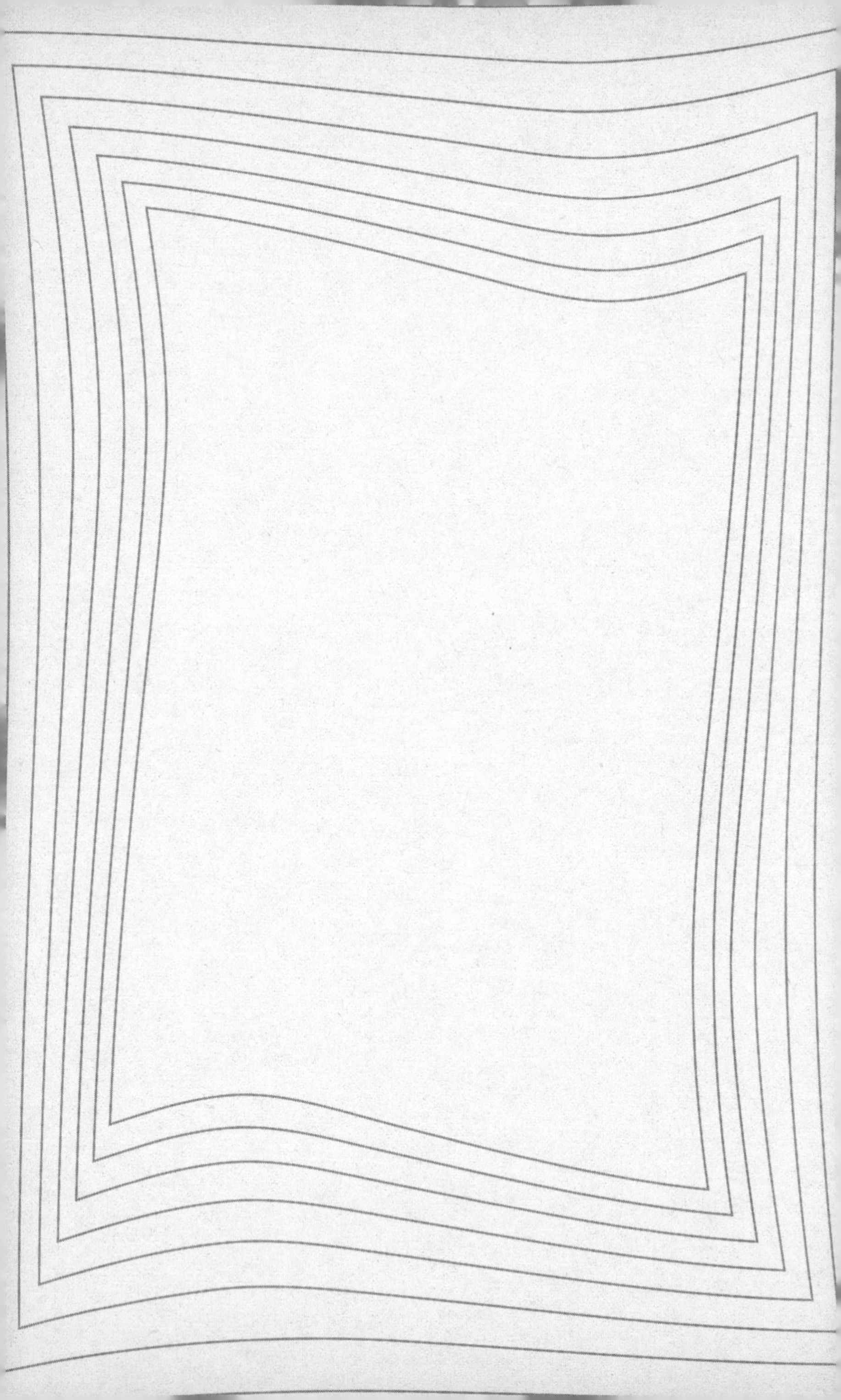

Part III

GLORY

Einstein and his second wife, Elsa Lowenthal, in Berlin, early 1920s

NINE

True or False?

EINSTEIN ALWAYS BELIEVED that there was an invisible framework to our universe, waiting to be found. He had always suspected, moreover, that this cosmic architecture would be very simple, and very exact, and very clear. And what could be simpler, or more exact, or clearer than an idea like G=T? It seemed impossible that his theory about space and gravitation could be false.

In the immediate wake of his breakthrough in November 1915, Einstein showed no hint of self-doubt—yet he knew that others had doubted him before. His earliest ideas about gravity, dating back to his thoughts in the Patent Office in 1907, had had only a limited impact. Even his initial elaborations, during his Prague years, had remained largely a private matter. But as Einstein's recognition among physicists had grown, resistance to his work in this area had grown as well. When he'd presented his extended theoretical elaborations at a conference in Vienna in 1913, seemingly the whole audience of distinguished faculty members had claimed that he must be deluded. At the time, Einstein had tried to stay calm, but later he admitted how shaken he had been. "My colleagues concerned themselves with my theory," he remembered, ". . . only with the intention of killing it dead." Even Max Planck, then Europe's most respected scientist, had had doubts, writing to Einstein, "As an older friend, I must advise against [publicizing this new theory] . . . You won't succeed, and no one will believe you."

Einstein knew he needed to convince his colleagues that his theory was legitimate, but perhaps most of all, he needed to reassure himself. Newton's theory of gravitation had been the bedrock of scientific thought for centuries. There was nothing in it about warped space. To one of his closest confidants, the Dutch theoretician Hendrik Lorentz, whom Einstein revered almost as a father figure, he admitted, "My business still has so many major hitches that my confidence . . . fluctuates."

Einstein was still fairly young, and had only recently received professional esteem. What he was attempting with G=T was terrifyingly bold. In essence, he was telling his colleagues that, like the inhabitants of Flatland, they had been blind to the fact that they existed within a higher dimension that they could not see. Now he claimed to have discovered it. No wonder they were skeptical.

What Einstein really needed was a test—some way of confirming the existence of this higher dimension around us. But how to pull out a test from something as seemingly abstract as the relation G=T?

He already had one possible way of proving his theory correct. He had been able to show, on the basis of his new equation, that the planet Mercury would advance in a fashion ever so slightly different from what Newton had predicted. The problem was that this wasn't really news; astronomers had already recognized that Mercury orbited differently than those predictions. Although no one—apart from Einstein—had been able to explain why this might be the case, cynical observers could always say that Einstein had started with those known facts about the orbit and worked backward to create a theory that would "produce" such an orbit.

What would be far more impressive would be if he could show that his new theory predicted something that nobody had imagined could possibly occur—and then go ahead, test that prediction, and show it was true. As early as his time in Prague, in 1912, Einstein had thought about this, and now he realized there might be a way to pull it off.

REMEMBER THE BALL bearing flicked forward on the taut trampoline. When it traveled where the trampoline was flat, it rolled forward in a quick straight line. When it got close to the sag at the center of the trampoline, where a small rock was bending it down—a rock that represented our sun—then the ball bearing veered inward as it dipped along that sag. Our sun is so massive that it produces a huge "sag" in space around it, and this is what the earth glides along, like a ball caught in a roulette wheel, with only the earlier forward motion preventing it from rolling even farther toward the sun.

In thinking about how to test his theory of gravitation, Einstein realized that according to his theory, it wasn't just the planets that would be pulled along by space's curvature in this way. Light, too, would be "bent" by gravity.

At first glance, that seems impossible. We're taught that if you shine a flashlight beam from one manned balloon to another, it shouldn't matter whether the balloons are high over the empty Pacific Ocean or whether they're floating right beside Mount Everest: the flashlight beam is going to travel in a straight line. It's not going to be tugged sideways simply because a massive mountain is beside it.

But the belief that light only travels in straight lines is a delusion, simply based on the fact that we live on a planet with weak gravity—or so Einstein suspected. If we could peer at realms where gravity was much stronger than it is on earth, we should, in fact, be able to detect invisible gullies opening up in space by seeing light swerve as it hurries along.

A variation of Einstein's simplest thought experiment shows how he arrived at this hypothesis. Instead of having a drugged explorer waking up floating freely in a closed room, imagine that you're the victim. And this time, instead of floating weightlessly, you feel a comfortable force pulling you down to the floor. Like the explorer's flotation, this force, too, is ambiguous. It could mean you've safely landed on the ground on earth, your terrifying journey is over, and when the airlock opens you'll get to step outside to a waiting, applauding crowd. But it also could mean that you are in a room out in space: one that has been hijacked by merciless marauders, who've attached a hook and are now tugging you forward to their evil mother ship. If their acceleration is calibrated right, you'll be pressed down to the floor with exactly the same intensity—no more, no less—than someone would feel in an elevator car waiting calmly for the door to open on the ground floor on earth. (This is an effect we recognize when we're in a car that suddenly accelerates, and we get pushed back against the seat. Close your eyes, ignore the roar of the engine, and you could be on a planet whose gravitational pull is tugging you into the seat just as hard.)

Suppose you're in the second scenario of Einstein's thought experiment—a hijacked, accelerating room, not a room resting on earth—and you manage to find a window, perhaps by lifting up a metal plate that had covered it. And suppose that, as you do so, the beam from a lighthouse on a convenient exuperant planet shines brightly into your window. If you weren't moving, you'd see the light beam come into the room and hit against the far wall, exactly opposite the window where it entered. But since your kidnappers are accelerating your own vessel upward, the light will not fall in exactly the same place it would if you were motionlessly suspended in space. Rather, in the time it takes for the light beam to move across your room, your vessel will have moved up a little, so that when the beam of light hits the far wall, it won't be exactly opposite the window where it came in, but instead will have curved to hit a little bit lower down.

This second part of the thought experiment reflects one of Einstein's driving views, what might be called observational democracy: the belief that just as no one automatically deserves superior rights in life, so no one observer can say that their vantage point in

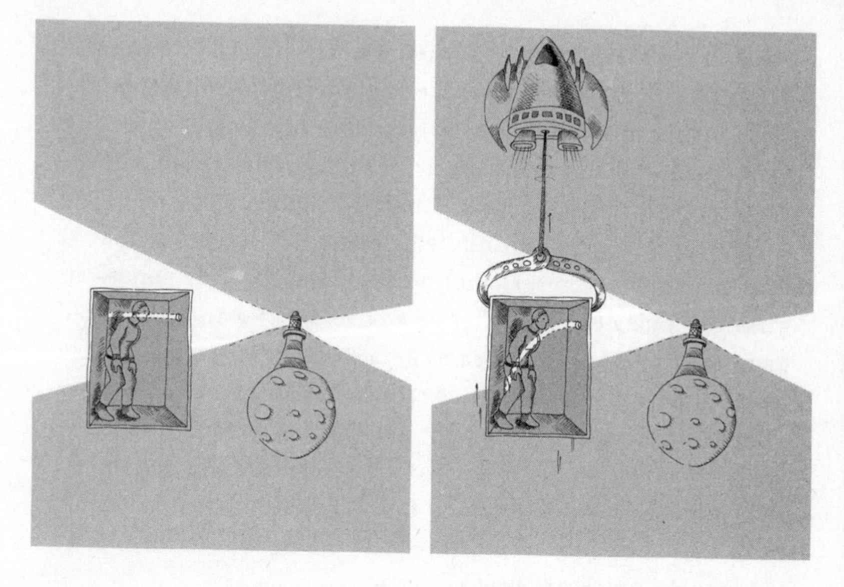

viewing some event is automatically superior to that of everyone else's. In the thought experiment at hand, what this means is that no one can possibly tell if they're being tugged in distant space, or if they're standing still in a closed room on earth—not if the tug pulling them along is of the right intensity. What someone sees inside one of those rooms will have to be exactly what they'd see if they were in the other one.

To see how this would hold true in the thought experiment, imagine how seeing the lighthouse from a stationary room on earth might compare to seeing the lighthouse from a hijacked room in space. In the tugged room, where you're held down to the floor with a force of 1 g (because the evil pirates are tugging you along), light bends as it travels across the room. In the static room on earth, where you're also held down to the floor with a force of 1 g (because the earth produces "real" gravity), light will also have to bend as it travels across the room. (Why? Because if the light bending wasn't the same, you'd be able to tell the difference between the two places, and that's what we've agreed is impossible.)

From this simple thought experiment, Einstein deduced that light bends in a gravitational field just as it would if viewed from an accelerating position. And that was the sort of prediction he could test. The rough approach had been in his mind early in his long years of work leading up to general relativity, although the details only became properly refined as he reached the final theory in 1915.

The real-life experiment Einstein envisioned was also simple, at least in scientific terms. He just needed to find a vastly ponderous mass, one bulky enough to create a huge sag in space near it, and then see if speeding light beams in fact veered off course as they passed close by, like a high-speed race car banking hard as it follows a curve. By observing the light beams that were visible around the periphery of such a massive object, Einstein predicted, one should be able to see what was behind it—all because the curvature of space caused by gravity would redirect the light from the hidden object to the observer's eyes.

In our solar system, Einstein realized, there was only one suitable candidate for such a test: the sun. It is so massive, and should bend space so significantly, that it should have a noticeable effect on the light around it. But there was one problem with this idea. Even if the sun did cause light near it to curve, Einstein knew, most of the time this would be too hard to detect. The effect would be quite small, just a fraction of a degree. During the day, when we can see the sun, its flames and explosions are so bright that they make it impossible to see the distant starlight that might be whipping close beside it.

But during a total eclipse? Then the best of both worlds combine. The sky is dark, but the sun is right overhead. Distant starlight that arrived in line with its edge would suddenly be visible. If that light had been bending, we would be able to see it.

Einstein had conceived of this test when he was first trying to elaborate the relationship between *G* and *T*, which he would finally pinpoint in his November 1915 breakthrough. But there was a reason he had been unable to report the results of such a test when he formally presented his theory of general relativity, with G=T at its

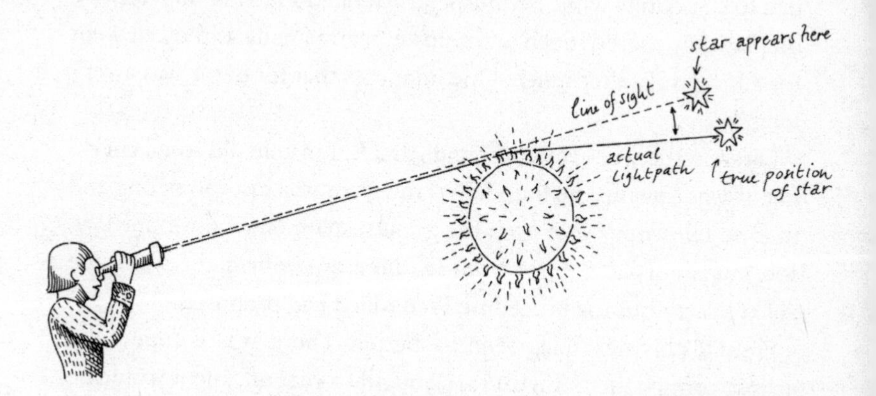

core, to Germany's greatest scientists late that year. Einstein had been entrusting the tests to an eager young astronomer named Erwin Freundlich, who had impressed him with his knowledge and keenness to help (and whose surname, appropriately, is German for "friendly"). But Freundlich turned out to be a man of stunning, probability-busting bad luck.

Freundlich's first suggestion to Einstein had been not to wait for an eclipse at all, but to go through old photographic plates stored at the Hamburg Observatory to see if those had inadvertently captured eclipses under the conditions Einstein required. Einstein wrote back with his support for the idea. Freundlich got permission from the observatory's director, started scanning and measuring plates, and found that although there were many—very many—plates in the observatory's archives, astronomers had repeatedly just missed recording the starlight deflections that would prove Einstein's theory true and win Freundlich respect and fame.

Freundlich was ever optimistic. Why not see if one could perhaps glimpse distant stars during the day and use measurements taken that way, rather than wait for an eclipse to darken the daytime sky? This was such an exciting idea that in 1913 he made a special trip to Zurich to talk it over with his new friend Einstein. Unfortunately, it was also Freundlich's honeymoon, which meant that his new wife had to sit politely while her husband attended a lecture on relativity, then as they shared lunch with Einstein, and finally as they all went for a long walk after lunch. One imagines that for her it was a very long day.

Later, in the weeks after Freundlich left, Einstein did some checking. It was obvious, he found, that the glare was just too strong and that no telescope—not even the great instruments on Mount Wilson in California, a letter from its director confirmed—would be able to carry out the procedure Freundlich had proposed.

Freundlich's next idea seemed better. There was a total solar eclipse coming up in August 1914, roughly a year off, and it would be visible not too far away, in the beautiful Crimea of southern Russia, near the sophisticated port city of Sebastopol. The Imperial Russian Battle Fleet had its headquarters there, which meant there would be restaurants and fine hotels nearby where he could celebrate when the imaging was done. At that time, Germany and Russia had been at peace for many years, so there was no reason to think anything could go wrong.

Freundlich's enthusiasm to confirm Einstein's initial ideas grated on some of the older researchers who constituted the astronomical establishment in Germany, and the official funding bodies resisted giving as much money as would be needed. Einstein couldn't believe how little belief they had in Freundlich. There wasn't much time to prepare! Before 1913 was out, he wrote Freundlich, "If the Academy won't play ball, then we'll get that little bit [of cash] from private quarters . . . If everything fails, I'll pay for the thing out of my own slight savings . . . [but] go ahead and order the plates . . . Don't let time slip by because of the money question."

Even with Einstein's help, there was a shortfall, but Freundlich managed to secure extra funds from the immensely wealthy Krupp family: the world-terrifying arms merchants. The Krupps' cannons and other military equipment were sold around the world, and they were the backbone of the German army.

Toward the end of July 1914, Erwin Freundlich arrived in the

Russian peninsula of the Crimea. A few days later, World War I began. Germany and Russia were on opposite sides. Freundlich was by then camping in the wilderness, outfitted with exceptionally powerful telescopes, not far from the headquarters of Russia's Imperial Battle Fleet. It's hard to imagine how one could make a German national seem more suspicious, especially as all his documentation showed that the Krupp Foundation was behind his mission. Freundlich's party was quickly surrounded by armed Russians, and his beautiful, lovingly prepared equipment was confiscated. When the eclipse took place as predicted, on August 21—the surrounding stars serenely distanced from the furious cannon flashes all over Europe—Freundlich was in a Russian prison camp.

Einstein and others managed to get Freundlich out of lockup, as part of a prisoner exchange not too much later. Characteristically, Freundlich wasn't downhearted. He simply had to engineer more opportunities to find fresh proof! The next really good solar eclipse wouldn't be for several years, which was too long to wait. But what if, instead of being so fixated on the sun, they thought of measuring starlight falling into one of the invisible gravity valleys that had to exist near the planet Jupiter? The deflection would be smaller than from starlight falling into the big curves in space near our sun (just as small pebbles on a trampoline surface make it sag less than heavier stones do). But Jupiter was certainly easier to photograph than the sun.

It wasn't a terrible idea, but by the time Freundlich began racing to get the right equipment together, his Hamburg Observatory director had had enough of this underling's exuberance. Einstein wrote to the Ministry of Education, encouraging administrators to bypass any bureaucrats who got in the way and give support to Freundlich. The minister passed the request on to the observatory director, who was not just a professor but a privy councilor who prided himself on being addressed by the title *Geheimrat,* or in a pinch "Your Excellency." He certainly didn't think of himself as a bureaucrat to be bypassed, and in his estimation Freundlich was a mere junior, of questionable competence and unacceptable insubordination. The director wrote a firm, and barbed, response to Einstein: "Even a 'multitude of the most sophisticated measurements' by expert observers, let alone by those who do not come under this heading, will not yield any useful results and merely cause a needless expenditure of time and effort."

The obstruction from Freundlich's boss was only one of the problems. As the war dragged on and Britain's naval blockade of Germany hardened, it proved impossible for Einstein and the loyal Freundlich to carry out their astronomical test. Einstein's bold new theory, it seemed, was dead in the water—that is, unless there was someone else to whom he could turn for help.

TEN

Totality

IN MAY 1919, a lean, sweaty Englishman stepped out of a hut on a small island off West Africa and looked up anxiously at the sun. A solar eclipse was coming, and he had spent two years readying for it. But if the threatening storm from the Congo coast didn't blow away, the expensive telescope he'd shipped down from England and lugged overland would be useless.

He had his team set them up anyway, despite the misty rain, and covered the lenses with his own jacket. It was a good thing, too, because suddenly, just a few minutes before totality, the clouds blew clear.

The sun's edge was shockingly bright. A previous generation of astronomers had imagined that somewhere within that glare, a fast-circling planet they'd named Vulcan existed. They had proposed that because something had seemed wrong with the orbit of Mercury. Newton's theory of gravitation predicted a very precise orbit for Mercury, but that didn't match what had been observed, even after correcting for the slight tugs the other planets in the solar system would exert on Mercury. An additional planet, orbiting closer to the sun, might be pulling Mercury into that uneven path.

Other telescopic surveys had failed to find this imagined new planet. If the large photographic plates ready to be installed in the Englishman's telescope showed what he expected, he would be able to disprove its existence beyond a doubt. He would do so not by documenting its absence on film, but by capturing evidence that would verify the theories of this still scarcely known German theorist, the gentle Berliner he'd never met, whose work had led to his presence on this distant island.

From his later records, we know what happened next. He gave a quick glance up. The clouds were coming back. There were going to be a lot of plates to change, and quickly. The Englishman bent back down, undeterred by the swarms of mosquitoes. He'd have time to speculate on the theory later, once he'd taken the pictures.

If, that is, the plates' emulsions could survive the tropical heat.

IN 1917 THE colleagues of the Cambridge astronomer Arthur Stanley Eddington were at an impasse. They knew that Eddington was a determined man, as those who tried to keep up with him on bicycle rides soon discovered. He always dressed properly, with neat suit trousers tucked in his equally neat socks, but with a nearly wild look on his face, he would speed through the countryside, ever faster, for hours on end, leaving his colleagues behind.

They knew that Eddington's determination came out in his religious views as well. Eddington was a devout Quaker, with principles that made him unwilling to defend the British Empire in the Great War that was still dragging on, so many years after it had begun. Many Cambridge men had died in the struggles on the Continent, including one of the university's greatest young physicists, Henry Moseley, meaninglessly facing Turkish machine gunners in the Gallipoli campaign. Eddington was showing signs of being one of his generation's leading astronomers. His colleagues at Cambridge were not going to have another of their own end up the same way.

When the Cambridge administrators tried to obtain an exemption from service for Eddington, however, by writing to the Home

Office, stating that the contribution he could make while remaining at the university was necessary for the war effort, everything had unraveled. The Home Office had written to Eddington with the appropriate exemption form. All he had to do was sign his name. Eddington conscientiously did that, but then even more conscientiously he added a postscript explaining that as a good Quaker, even if he wasn't deferred on the stated ground, he would claim it on the ground of conscientious objection anyway. As one friend of Eddington later put it, "This postscript naturally placed the Home Office in a logical quandary, since a confessed conscientious objector must be sent to a [prison] camp." Eddington's colleagues were "very much piqued."

Luckily for him, and for Einstein, Eddington's friends landed on a solution that didn't involve war or a prison camp. Rather, it involved Britain's enemy Germany and the strange scientific theories that had been leaking out of that country even at the peak of the war.

Direct contact with German scientists had closed down since the war began. Censors didn't like telegrams with obscure formulas and lists of numbers traveling between the two countries. There was also a general distaste in England for all things German, leading at times to riots and even to some anxious immigrant families changing their last names. But hints of Einstein's new ideas had made their way to England through trusted intermediaries in the Netherlands.

Eddington's main protector, Astronomer Royal Sir Frank Dyson, couldn't understand all the details of Einstein's theories—and wasn't even convinced they were necessarily valid—but he did recognize what a coup it would be if a Cambridge man could find out for sure whether this strange German scientist was right. Not only would Eddington be able to demonstrate that science could transcend the barbarity of warfare; he'd also be able to preserve a few precious links between his nation and Einstein's.

Dyson spoke with his contacts at the Admiralty and came away with an arrangement so securely signed and sealed that even an ear-

Eddington as a young man, c. 1914

nest Quaker couldn't get around it. Eddington would be engaged in important government business and under no circumstances could be sent to the deadly front, or even relegated to a prison camp. Instead, Eddington would be involuntarily volunteered to run an astronomical expedition to test Einstein's theories, once and for all.

The idea of a scientific, rather than a military, mission was fine with Eddington. Perhaps even more than Dyson, he appreciated the salutary effect science could have during wartime. One of Eddington's noted contemporaries in England's Quaker community, Ruth Fry, wrote that "one person who heads an expedition to heal the wounds and desolation of war is stronger than a battalion of men under arms." For Eddington, a voyage to promote the views of a thinker working in the capital of Britain's gravest enemy would be perfect. "The lines of latitude and longitude pay no regard to national boundaries," Eddington wrote. The hunt for truth would bring mankind together.

And so, in the midst of wartime Britain—with shortages of almost all materials and the seas around the island nation stalked by deadly German U-boats—Eddington began to plan how to succeed where Einstein's helper, the unfortunate German astronomer Freundlich, had failed.

Eddington knew that an eclipse was expected on May 29, 1919, so he determined to use it to test Einstein's theory. Eclipses are only visible from specific regions, and this one was expected to follow a trajectory across the Atlantic from northern Brazil to Africa. Eddington and Dyson arranged for two teams, one to view the eclipse from the town of Sobral, in the Brazilian jungle, and the other to try to reach the island of Principe, a Portuguese colony just off the coast of West Africa, right near the equator and along the course of the eclipse.

No one—not even the shipping insurers at Lloyd's of London—had any account of steamers that went to Principe, so the second team would have to get as close as they could and then hope they could work out the details from there. Complicating things further, Eddington's limited funds would only allow four Englishmen to go on the expeditions that he had planned: two of Eddington's colleagues would observe the eclipse from Brazil, and Eddington himself plus a mechanically gifted technician from the Greenwich Observatory he'd chosen, Mr. E. T. Cottingham, would watch it from Principe.

Experimentalists from another country were sometimes brought along on such expeditions to help researchers from the sponsoring country, but in this instance one obvious foreign candidate was left out. While the war was still on, it would have been impossible for Freundlich to accompany either team, and even after the Armistice was signed in November 1918, such transnational collaboration proved to be a bridge too far for the poor man. Freundlich no doubt knew that May 1919 was his one big chance, for no other eclipse where the sun passed through such a dense star field would occur for many years. Because most communications between the two sides were still blocked, he might have continued to hope that an invitation would arrive. He spoke English well enough and could get Einstein's own recommendation. But by February 1919, the month the expeditions were scheduled to depart, he'd have known that wasn't going to happen and he would be left behind.

IN ENGLAND, PREPARATIONS for the two expeditions went distressingly slowly at first but picked up pace once the war wound down. "It was impossible to get any work done by instrument-makers until after the armistice," Eddington wrote. Once fighting did end, in November, they only had three months to get ready. Just before the British team left England, an astronomer who'd originally been written in for the Brazil leg but couldn't make it—one Father A. L. Cortie—suggested that along with their main equipment, the men also take a relatively small four-inch telescope as a backup in case something went wrong. Eddington already had a lot to pack, but Cortie was insistent, and so the telescope ended up with the luggage of the team going to Brazil.

In February 1919, with telescopes, crates, canvas, mirrors, cigarettes, two metronomes, no doubt plenty of tea, and other essential items securely packed, the four men gathered in the port of Liverpool. They found there the *Anselm,* a ship at their beck and call, which would prove particularly well suited for crossing a dangerous sea only recently cleared of German U-boats. They departed England on March 8, 1919.

At Madeira, the Portuguese island off the coast of Morocco, they split up, with the Brazil team continuing on, while the pair bound for Principe stayed ashore, as Eddington hunted for a ship they could get space on the rest of the way. That took nearly a month. Cottingham was bored, but Eddington, though sadly without his bicycle, used his time to scale the local mountains, and also to visit Madeira's casino—writing to his mother that this was not for any gambling, but simply because word was out that they served quite good tea there. If Eddington had chosen to gamble, his mathematical quickness would likely have added considerably to the expedition's coffers.

Eventually, in early April, he found a transport that would take them down to the tropics. The world was only slowly recovering from the war, and as they left the harbor, they passed sunken ships with twisted metal masts leaning out of the water. On the open sea, passengers weren't informed of their daily location, for despite the Armistice, no peace treaty with Germany had yet been signed, and officially a state of war still existed.

Dyson hadn't entirely understood Einstein's new idea, but he knew enough spherical geometry to be able to chart, back in his rooms at Greenwich, the approximate route that Eddington and Cottingham would have to follow. This, too, demonstrated what geometers had learned. If the earth could be opened up, a straight line from Madeira to Principe would, of course, be a far shorter path. But since that was impossible, they'd have to take the longer route, along the curved surface of the planet.

Eddington knew that, too, but since the earth is so big, from his position on the ship, so close to the surface, the horizon always seemed to be straight ahead, oscillating only from the waves that lifted and dropped them. There was the smell of burnt fuel as the engines pushed them forward, and the tedium went on for day after day until—from his private notebook—"on the morning of April 23, we got our first sight of Principe."

The island shot out of the sea, with central mountains a half mile high seeming to drag heavy masses of cloud over them. There were dense forests everywhere. In places wild surf hit the bases of cliffs that climbed up five hundred feet, but there also were coves where the ocean had worn away the volcanic rock, and in one of these the researchers put ashore.

At about 80 degrees Fahrenheit, Principe wasn't as hot as one might expect the equatorial tropics to be, but it was humid, and they'd arrived just before the end of the rainy season, so tremendous storms were still constant. In between storms, the island was rich with clouds of mosquitoes. Eddington and Cottingham had to cover themselves, despite the heat, to avoid being bitten to distraction. They took quinine daily, had local laborers build huts that were at least partly waterproof, and chased monkeys away, sometimes with rifles. A more poignant reminder of how far from home they were came when one of the island's plantation owners invited them over and casually put out full bowls of sugar. They did something of a double take: because of wartime rationing, they had scarcely seen sugar for five years.

A little over three weeks after they arrived, it was time to prepare for the eclipse. The worst of the rains had stopped several days before, and to be extra sure of avoiding clouds, they had moved as far from the central mountains as possible, to a plateau on the northwest tip of the island. The violent Atlantic lay a steep several hundred feet below. The forest was so thick that their equipment couldn't be transported by mules for the last kilometer, and only native porters could help them. They found a clearing, and from there, on April 29, they finally were in position for what they had come for.

Eddington recorded the beginning of the eclipse in his journal, calmly recalling the morning's meteorological phenomena. "[In] the morning there was a very heavy thunderstorm from about 10 a.m. to 11.30 a.m.—a remarkable occurrence at that time of year." Then the sun appeared, but only briefly before the clouds rolled back in. As the day proceeded they had tantalizing glimpses of the sun, and by 2 p.m. the drifting clouds only lightly covered it.

There would be no more than five minutes of totality, and that was going to start precisely at five seconds after 2:13 p.m. Eddington must have been eager for the obstructing clouds to drift away quickly. If Einstein was right, the sun was already distorting the space overhead—like the rock on our taut trampoline—so that light from the stars in the distant Hyades cluster would bank hard as it took that curve. The starlight would have been traveling for trillions of miles by that point. Yet if it were blocked by clouds just a few hundred feet above Eddington's telescope, he would never be able to prove anything.

Cottingham had the all-important metronome ready and began to give warning to Eddington at about 58 seconds, 22 seconds, and 12 seconds before totality. When the last visible crescent of the sun disappeared and the forest beyond their clearing fell into near-complete darkness, he called out the single word "Go!" Eddington had been holding the first photographic plate and now quickly slotted it in, as gently as possible so as not to jar the telescope. Cottingham kept on counting, calling out every tenth or twentieth beat, so that Eddington would know when to pull out each plate to ensure that the exposure was right.

It was a nerve-racking five minutes, and when it was over, the group was in a somber mood. As Eddington recalled, "We had to carry out our programme of photographs in faith." Since he had to keep on changing plates, he scarcely looked up to see the eclipse. Halfway through he did take a quick glimpse to estimate the cloud cover. By the end they had taken sixteen photographs, but since there had been so much cloud they didn't know if any of their pictures would be usable. Everyone was disappointed. Then, to make it worse, only minutes after the peak moments of the eclipse the sky cleared entirely.

From that point on the researchers were consumed with decoding the photographs. They were able to develop two each night, and they did that for six nights, while during the daytime they started trying to measure the distant star displacements they were looking for. But the spotty results they had due to all the clouds meant that Eddington couldn't yet be sure if what they'd recorded confirmed Einstein's predictions or not.

The best Eddington could conclude was what he wrote in a telegram he left on Principe to be sent to Dyson: THROUGH CLOUDS STOP HOPEFUL STOP EDDINGTON. Before he could complete the careful measurement of the displacements—effects that would appear as movements of bare fractions of a millimeter on his plates, scarcely more than the width of a human hair—they had to get off the island. One of the planters had told them of rumors that there was going to be a steamship strike, and Eddington decided that he had better take the first boat back, since otherwise his team might be stuck on the island for months. The sea journey could damage the developed plates, but they had been away from Cambridge long enough.

If Eddington was chagrined about the outcome of his study after he returned to England, he could at least comfort himself that his team wasn't the only one that had struggled with their measurements. The Brazil expedition staggered back later, and they'd suffered an even greater disappointment with their big telescope. The sky had been clear enough, and conditions had been much better than on the rugged island of Principe. They'd had use of what was apparently the first car ever seen in that region of Brazil to haul their equipment, which they neatly set up on the conveniently flat racecourse of Sobral's Jockey Club. There was cool, if not quite cold, water available to develop their test plates. In the days before the twenty-ninth, interested locals even lined up to buy tickets to look through the telescope.

But the very fact that the sky was so clear proved to be a problem. The Brazilian team was barely four degrees away from the equator, and the direct heat distorted their primary apparatus. The team's notes, scribbled down as they were developing the plates the night after the exposure, record their foreboding that their observations might have been a failure: "3 am . . . There has been a serious change of focus, so that, while the stars are shown, the definition is spoilt." They realized this was because the intense daylight heat had made their telescope's mirror expand unevenly.

The main telescope had failed the Brazilian team—but Father Cortie had known what he was doing when he'd insisted on sending the extra four-inch telescope along. From a sense of obligation as much as anything else, the Brazil team had inserted a handful of extra plates at the ideal focal point of that small device—and from that had ended up with the best plates of the entire expedition: better than those from the heavy telescope at the Jockey Club; better than Eddington's equally large telescope arduously transported so high above the Atlantic on the raw cliffs of Principe.

When Eddington and his assistants in Cambridge analyzed the plates from Principe, they worked separately to be sure their individual handling didn't affect the readings. Two of the plates weren't quite as bad as they'd feared, and Eddington could incorporate their results as well. As they worked, they knew that Einstein in his final calculations in 1915 had come up with an estimate for light streaming from a distant star, and it would be bent only a very small amount. Hold your little finger at arm's length, and its width is about 1 degree of arc. Astronomers divide that degree into 60 minutes, and each of those minutes into 60 seconds. Einstein's prediction was that incoming starlight would be diverted a scant 1.7 seconds of arc (symbolized as 1.70") as it passed near the sun, compared to where it would be if the sun wasn't there and the space it traveled through was flat. That's smaller than the slightest scratch you can see on your finger. Such measurements are hard to detect. Would they conform to Einstein's predictions, or would they sink his brave new theory once and for all?

DYSON AND EDDINGTON had a great sense of the dramatic and planned to hold off announcing the results until they could assemble a large, distinguished audience. This delay also meant that scientists who'd heard rumors of what was going on became anxious to learn what had really happened. From Berlin, Einstein—who later pretended to have known all along that he would be proved right—ever so casually wrote to a physicist friend in the Netherlands, asking, "Have you by any chance heard anything over there about the English solar eclipse observation?"

In November 1919, six months after the eclipse, Eddington was ready. The findings would be presented at a grand joint session of the Royal Society and the Royal Astronomical Society, in the august setting of Burlington House, the mansion where both were headquartered, on Piccadilly in London. Depending on their findings, the world would learn whether Newton's theories—which had dominated all scientific thought for more than two centuries—were to be overthrown, or if the bizarre predictions of the Swiss/German theorist Einstein were worth no further attention. The fact that Newton had once served as president of the Royal Society and that his presence was still very much felt within its ranks only raised the stakes.

Tea was served at 4 p.m., as always, and in proper English style the guests had to pretend they had no special interest in what was going to happen next. Finally, at around 4:30, it was time to begin. Frank Dyson strode to the podium. The philosopher Alfred North Whitehead was in attendance and later recalled, "The whole atmosphere of tense interest was exactly like that of the Greek drama . . . There was a dramatic quality in the very staging:—the traditional ceremonial, and in the background the picture of Newton to remind us that the greatest of scientific generalizations was now, after more than two centuries, to receive its first modification. Nor was the personal interest wanting: a great adventure in thought had at length come safe to shore."

Dyson spoke, and then the head of the Brazil expedition spoke, and finally it was Eddington's turn to announce the expeditions' results. Over a year's work had been building up to this moment, and much of Einstein's efforts hinged on it as well.

Had he been in the room, Einstein would not have been disappointed. The predicted deflection, Eddington announced, was 1.70". The most trustworthy results from the two expeditions came out at 1.60", with a margin of error of 0.15". Dyson said it simply: "After a careful study of the plates I am prepared to say that there can be no doubt that they confirm Einstein's prediction"—his prediction, that is, that light would curve when it got close to the sun. Based on the latest scientific evidence, Einstein's new, geometric picture of sufficiently massive things curving space enough for us to detect had been shown to be true.

One unconvinced member of the audience pointed to the portrait of Newton and said, "We owe it to that great man to proceed very carefully in modifying or retouching his Law of Gravitation." No one was listening. In fact, the official chair of the meeting—the elderly Nobel laureate J. J. Thomson, discoverer of the electron—stood up to finish and put his respected word on Einstein's side. "This is the most important result obtained in connection with the theory of gravitation since Newton's day," he told the crowd. "It is . . . the result of one of the highest achievements in human thought."

The thinker behind this "highest achievement" was still unknown to the general public, but the scientific establishment had given his theory the ultimate official support. It wouldn't be long now before the world knew the name Albert Einstein.

INTERLUDE 2

The Future, and the Past

BACK AT CAMBRIDGE, over a decade after his fateful expedition to Principe, Arthur Eddington found himself sitting before the fire in the Senior Combination Room at Trinity College, along with Ernest Rutherford, the director of Cambridge's greatest physics lab, and a handful of other guests. The topic of fame came up, of public celebrity, and one young guest asked why in the previous few years Einstein had had so much public acclaim, while hardly anyone among the general public knew who Rutherford was, despite his Nobel Prize. After all, it was Rutherford, more than anyone, who'd uncovered the inner structure of the atom.

"Well, it's your fault, Eddington," Rutherford teased. Not everyone immediately grasped his meaning. All present—including the bright young Indian researcher who would later relate the story—knew that Eddington's dramatic presentation at the Royal Society in November 1919 had had some effect on Einstein's reputation, but why had it been so overwhelming?

The men settled into their deep chairs, and Rutherford spoke, more reflectively this time. The war had just ended when Eddington announced the results of his study, Rutherford recalled. Astronomy had always appealed to the public imagination. Now people learned that an astronomical prediction by a German scientist had been confirmed by British expeditions—prepared while the two countries were at war—to Brazil and West Africa. Harmony was possible. True peace was possible. The discovery "struck a responsive chord," Rutherford concluded, "and then the typhoon of publicity crossed the Atlantic."

And a "typhoon" it was, for what had happened to Einstein after that Royal Society meeting was unprecedented—unimaginable, even, at least at the time.

It began, as many things do nowadays, in the media. The London *Times* had been moderately restrained in its coverage of the meeting, but the same could not be said for its many counterparts across the pond. Although the *New York Times* had some excellent reporters, the best one it could get to London for the Burlington House meeting on short notice was Henry Crouch, the paper's main golf correspondent, who'd thought he was going to be spending his time in Britain at St. Andrews and similar beguiling links. He would have been the first to admit that he was very much not an authority on the mathematics of four-dimensional space-time. Crouch did, however, work out that something extraordinary had occurred, and his enthusiasm was transmitted to the *New York Times'* headline writers. Hence, just six days after the big meeting, the newspaper reported:

LIGHTS ALL ASKEW IN THE HEAVENS
Men of Science More or Less Agog Over Results
of Eclipse Observations.

EINSTEIN THEORY TRIUMPHS
Stars Not Where They Seemed or Were Calculated
to be, but Nobody Need Worry.

A BOOK FOR 12 WISE MEN
No More in All the World Could Comprehend It, Said Einstein
When His Daring Publishers Accepted It.

The headline was appropriately breathless but impressively incorrect. The stars were exactly where Einstein had predicted: that was, indeed, the whole point of the expedition. Crouch had never spoken to Einstein and had made up the quote about only a dozen men being able to understand the theory.

None of that mattered. Rutherford was right about people liking the international harmony that Eddington's expedition had demonstrated. There had been some other examples of coorperation after the war—in exploration and medicine—yet Einstein alone was given an open-top-car parade before tens of thousands in the United States, saw massive lecture halls filled hours before he appeared in Prague and Vienna, and was mobbed at movie premieres. When he was home in Berlin, letters constantly arrived, hundreds of them, and then thousands. They were delivered in such volumes that Einstein once had a dream in which he couldn't breathe, for "the postman was roaring at me, hurling bundles of letters."

It helped that Einstein had an informality that contrasted with the snobbery of the upper classes who had led the world during the Great War. Reporters loved the fact that once, when he was arriving to give a grand address at the University of Vienna, officials at the train station waited, and waited, for the great man to emerge from the first-class compartment. Then—shades of Max von Laue's visit to the Patent Office in 1907—they saw a familiar shape, far down the platform, walking contentedly along on his own from the third-class car he had taken: violin case in one hand, briar pipe and suitcase in the other.

But there were further reasons for the fame. Looking upward to the stars can be felt to be the same as looking upward to the divine. Mankind had always wanted to understand the ways of God—to know why chaos occurs and how the meanings we want to believe lie behind it could be found. And this, the world was convinced, was what one quiet, thoughtful Swiss/German physicist had discovered.

Most of all, however, Einstein's fame was a result of the trauma the world had just endured. Millions of men had died in the Great War, and innumerable families had lost a father, a son, a husband. There was a need to find some way back. Séances became popular, even though they were repeatedly shown to be run by charlatans. It was too painful to think that the dead were so completely gone that no contact—not even a whisper—could be maintained. An alternative seemed more plausible than it might have in earlier times, for homeowners were beginning to install large electrically operated machines—the first radios—in kitchens and living rooms, and through these devices, one could hear voices that had traveled invisibly over long distances. Who knew what else might be invisibly traveling, waiting, somewhere beyond?

This is what Einstein's work also seemed to promise, for he showed that at least some forms of time travel are definitely possible. Before Einstein, it was taken for granted that we live in three dimensions and that, quite separately

—at right angles to that, one might say—there is a fourth dimension, of time, which we move through at a steady, unchanging forward rate. Einstein transformed all this. The reasoning that led to his prediction of starlight bending as it neared the sun also leads to this prediction that time "bends" depending on how strong gravity is around it. Usually we don't notice this, because the effects are very slight in the fairly weak and uniform gravity around us on earth and at the speeds—so much less than the speed of light—with which we move about. But Einstein had unveiled this unsuspected truth about time, and with the success of Eddington's expedition, everyone now learned that he was right. In particular circumstances, some of us can travel forward through time—can be sped into the future—at greater rates than others.

Strange implications follow from the facts of nature that Einstein discovered. Think what might happen when our explorer who was kidnapped by space pirates and dragged at high speeds through the galaxy is finally rescued. The explorer has been living within time that from his perspective is moving more slowly than that of his rescuers; the rescuers have been living within time that from their perspective is moving more quickly than his. Of course, if they manage to free him from the space pirates without much delay, there will be little chance for that difference to build up. But if the pirates haul him around on a huge roundabout journey before the rescuers finally reach him, they might be decades older, while he—undergoing sufficient acceleration—will have aged only a few days. If his journey has seen truly tremendous acceleration, he might be just a week older when he's found, but the original rescuers will have long since died, and it will be their distant descendants who greet him.

This mind-bending stuff isn't merely imagined, postulated, unproven. Einstein showed that the effect is not just on our measuring machinery, but in reality itself. A traveler to the stars could return after what to him was genuinely no more than two or three years, but while he would still be a young man, millennia would have passed on earth, and everyone he knew—possibly even the civilization he'd left—would have long since vanished.

Were these effects to be magnified so that they were noticeable even at the ordinary speeds and in the usual gravitational fields we're accustomed to on earth, someone driving fast enough to an exercise class would be in the car for only one minute by his measurement, while his friends waiting for him there would be watching him drive for half an hour of their time. Parents who could afford to rent apartments on the very top floors of tall skyscrapers—where gravity is weaker—would age much more slowly than children they'd left in boarding schools on the ground. They could pass a single week up there while their children trundled through all the years from primary school to graduation.

These were the sorts of results that prompted such baffled comments as that from the distinguished scientist and Zionist leader Chaim Weizmann: "Einstein explained his theory of relativity to me for weeks, and by the end I was convinced that he understood it." But Eddington's findings showed that somehow, extraordinarily, relativity was true. Distant starlight didn't veer around the sun only because space itself was sagging. Rather, time was operating at different rates as well. (This is hard to envisage, but imagine incoming starlight being made up of a row of light beams all rushing forward in parallel, like a row of sprinters. The ones on the outside get a longer time to advance a given distance, and so, like the runners taking a banked curve, that's why the entire row begins to swerve.)

How much further could Einstein's insights go? The fact that with the right technology, we could accelerate into the future was impressive. But after the Great War, many people would have given anything to be able to travel in the other direction, to the past—if not to bring back lost life, then to get more time, even if just one final hour, with those they loved before a bullet or a shell brought them down.

Although some recent developments of Einstein's work have suggested that it may in fact be possible to travel backward in time, in the immediate aftermath of Eddington's expedition, no physicists, not even Einstein, saw how to do that. But they did, even at that time, appreciate another solace-giving implication of his theory: not quite the ability to travel into the past, but not quite an acceptance that those we love are entirely lost either.

In the world before Einstein, everyone believed that of course two events that one person finds simultaneous have to be just as simultaneous for everyone else. But Einstein's work revealed that this is not so. Even several years after World War I ended by our reckoning of time, there were locations beyond our galaxy from which the vast numbers of deaths in the trenches and on other battlefields had not yet occurred.

This was not just an artifact of measurement or a fantasy of the mystics, as with William Blake's "I see the Past, Present, and Future, existing all at once/Before me." If we could be in one of those far-off locations right now, we, too, would be living at a time when a friend or husband who had been shot was still alive. The catch, however, was that those perspectives involved such tremendous speeds and relative accelerations that, Einstein's equations showed, we would never be able to access them, since we could never travel fast enough to get there: present-day technologies could bring us nowhere near the required speeds.

Still, the knowledge that such realities were possible, even if only on a theoretical level, gave comfort to many—including Einstein himself. Many years later, when his friend Michele Besso died and he himself was seventy-six—with heart and other health problems, and knowing his own end was near—he wrote to Besso's family of the deep understanding he drew from this view: "Now he has preceded me briefly in his departure from this strange world. This means nothing. For those of us who believe in physics, the distinction between past, present, and future is only an illusion, however tenacious this illusion may be."

Despite—or more likely because of—its ballooning popular appeal, much of Einstein's theoretical work became distorted in popular understanding. Almost immediately after Eddington's results were publicized, books, lectures, and radio broadcasts went out about the great man's work, many of which got his theories wrong. But enough of what Einstein had achieved made its way through.

No other scientist had ever been so acclaimed, even if no one—not Rutherford, not Einstein himself—could be sure why that was the case. Yet whatever the reasons, practically overnight, multitudes of people came to think of Einstein as someone who had seen what mankind had never imagined—who had reached into the heavens and brought down, if not salvation, at least a glimpse of what deeper reality there might be.

ELEVEN

Cracks in the Foundation

EINSTEIN SHOULD HAVE been happy. Revered worldwide since Eddington's confirmation of his theory in 1919, he was awarded the Nobel Prize of 1921 for his work in theoretical physics. Movie stars and royalty wanted to be near him; the mobbed appearances continued. But amidst that acclaim, amidst that fame, Einstein began to worry about one consequence of his celebrated theory—and his professional angst was also compounded by growing stress in his personal life.

His divorce from Mileva Marić (which had finally come through in 1919) had given him freedom, but it had distanced him from his two beloved sons. He tried writing them long chatty letters, but they were in no mood to accept their father's overtures. When he got them to visit him in Berlin, he purchased a telescope and put it on his balcony for them to use, but this didn't help either. When Einstein did travel to Switzerland to take them on the sort of walking holidays they had liked before, everything was mannered, stilted. Once, in exasperation, he wrote to the elder boy, Hans Albert, from Berlin, taking him to task for being so cold. But Hans Albert was just as angry: his father was abandoning them, so how could he expect any kindness in return? Hans Albert later remembered that he felt as if a "gloomy veil" had come over what was left of their family life.

Einstein raged at Marić for poisoning his children's minds against him, but he must have known that he was partly responsible—and for what? Life with Elsa Lowenthal hadn't worked out as he had hoped. He had intended to keep the liaison strictly on his terms, having written to Besso in 1915 that it was "[an] excellent and truly enjoyable relationship . . . ; its stability will be guaranteed by the avoidance of marriage." Lowenthal, however, had a different view, and in June 1919—while Eddington was still on the tropical island of Principe—they had married. Almost immediately after the wedding, something changed. Marić may have been resentful of the way she was left out of his scientific discussions, but at least she had understood the main lines of his work. Yet although Lowenthal's lack of scientific education had been fine when Einstein was on the rebound, now he was discovering that behind her natural ebullience lay an intellect that left much to be desired. "She is no mental brainstorm," he later remarked.

During their courtship, Lowenthal had agreed with Einstein about the pleasures of an informal life and had enjoyed his mocking of wealthy, established Berliners. But once they moved into her seven-room apartment in a building with a grand lobby and a uniformed doorman, he felt trapped among her Persian carpets, heavy furniture, and display cabinets filled with fine porcelain. Some of her friends were thoughtful, but the majority, he was coming to see, were just chattering socialites. Worst of all, she began babying him. "I recall," her daughter wrote, "that my mother often said during lunch, 'Albert, eat: don't dream!'" It was all very far from romantic.

Soon Einstein began to have affairs. His mere presence, an architect who knew him well remembered, "acted upon women as a magnet acts on iron filings." Some of these women were younger than Elsa, some richer, and some both. What they saw was one of the most famous men on the planet, yet one who was unlike the stereotype of the desiccated intellectual. He was still fit and broad-shouldered (as friends who saw him take off his shirt noted); he loved telling wry Jewish jokes, and he had a direct, Swabian use of language. Actresses such as the renowned Luise Rainer soon wished to be seen with him. He spent evenings with a wealthy widow at her villa in Berlin and accompanied another woman, a fashionable entrepreneur, to concerts or the theater, riding with her in her chauffeured limousine.

Einstein with the German actress Luise Rainer, mid-1930s. Her husband was jealous of her flirting with the great scientist, although the peak of his philandering had actually occurred a decade earlier.

The contrast between these other women and Elsa, with her chatter and her increasingly baffled disappointment, was painful for everyone. Einstein liked to go sailing, and when he did manage to find free time would head to their country house near a lake not far from Berlin, where he kept his sailboat *Tümmler* (German for "porpoise"). He would go out alone in the boat for hours, dreamingly adjusting the tiller as the winds skidded him here and there. His housekeeper described one regular visitor to the summer house when Elsa was away. "The Austrian woman was younger than Frau Professor," the maid recalled, "and was very attractive, lively, and liked to laugh a lot, just like the Professor." On one memorable occasion, Elsa found another woman's "article of clothing" still on the boat, and they had an argument that, in its cold fury, continued for weeks. Men and women were not designed to be monogamous, he insisted. Elsa confided to a few close friends that living with a genius was not easy—not easy at all.

This was not the marriage either of them had wished for. In the letter Einstein wrote to Besso's adult children, consoling them after their father's death, he concluded: "What I admired most in him as a person was the fact that he managed for many years to live with his wife not only in peace but in continuing harmony—something in which I have rather shamefully failed twice."

If this were Einstein's only failure, it might have been bearable. But he was confronting an even worse problem. Even as early as 1917, at what should have been the peak of his accomplishment, Einstein had discovered what seemed to be a catastrophic flaw in his great G=T equation, and it had been preying on him ever more as the 1920s went on.

IMMEDIATELY AFTER COMING up with the equation that explained gravitation in December 1915, Einstein had been jubilant but exhausted. Only as 1916 went on did he begin any other work, and only by the end of 1916 did he have the energy to return to G=T.

Everything he'd done so far with that equation had focused on how it applied to particular objects, such as the orbit of Mercury in our solar system, or the path of light from particular distant stars as it traveled near our sun. Now, he decided, "[I] wish to take larger portions of the physical universe into consideration." The idea was to explore how G=T might apply to the mass of the entire universe.

This is when Einstein found what seemed to be a catastrophic flaw. The scientists of his era believed that the universe was static, fixed, unchanging: filled with a collection of stars, stretching away to a very great distance, some of which might slightly move from place to place but which, overall, never changed at all. Yet G=T predicted something quite different. If the "things" floating in space were already separated enough from one another, his equation allowed their random motion to start sending them even farther away from one another. But worse, his equation also appeared to allow another possible scenario. If a certain number of the "things" floating around in space were close enough so that they did start clustering together, the curvature in space that created might make even more objects start sliding toward them, thereby producing a runaway collapse.

The effect would be as if an enormous object landed in the Pacific Ocean and generated such a great whirlpool that everything on the planet—water, then islands, and soon entire continents—started being sucked toward it. The equivalent on the scale of our universe would be a sky-spanning "valley" taking shape in space, making everything tumble into it. Even more, the valley would start folding in on itself as the density of all the things accumulating in it—all the mass and energy that fell in—made the geometrical curve ever greater, as space itself began to collapse.

Einstein wasn't an astronomer, but he did know the basics—enough to believe that the scenario his theory was generating seemed impossible. Our solar system has planets that spin around a single, central sun. Our Milky Way galaxy is full of similar stars: some bigger, some smaller, but all, it was believed, hovering in fairly fixed positions. That's all there was. It was what the philosopher Immanuel Kant had described as an "island universe": fixed, stable, unchanging for all time. That's why the constellations the ancients had spoken of—Virgo, Sagittarius, and the like—were still roughly in the same positions in the night sky. But Einstein now saw that if his simple G=T equation of 1915 was true, that couldn't be the case, and everything would constantly move.

Thus his dilemma. He loved his equation's simplicity and clarity. It was wonderful to think the universe was arranged to follow such a simple, beautiful law. It made exciting, crisp predictions about what was happening within our solar system, as with starlight veering off course near the sun. Yet his equation also seemed to predict that on a far larger scale, the universe as a whole was changing—that all the stars in the heavens would one day either fly away forever or start falling together in a giant collapse. Every respected astronomer, however, was insistent that was false, for all their observations seemed to show that the universe was fixed, stable, forever unchanging in size. How could the consensus of all the world's top astronomers be wrong?

Something had to give, Einstein decided, and if the observable facts about the universe wouldn't change, he'd have to. Since his 1915 equation predicted that the universe was changing, he had to fix the equation so that it wouldn't make that prediction. What it said about small-scale effects, such as our sun making space sag enough to deflect starlight passing nearby, would still be allowed to hold. But what it said about larger-scale effects—those shaping the universe as a whole—would have to be corrected. In February 1917, accordingly, addressing the Prussian Academy in Berlin, Einstein declared, "The fact is, I have come to the conclusion that the equations of gravitation hitherto presented by me need to be modified, in order to avoid these fundamental difficulties."

He'd need to change his beautiful G=T equation, but how?

Einstein had mulled over the problem at length, and in his 1917 address he presented the only possible patch he could think of. He would have to insert an extra term in his original equation. This new term would take away some of the power on the left-hand side of the equation—the one concerned with the geometry of space. It would come to be known as the cosmological constant, because it was a fixed, or constant, number that operated on the level of the cosmos. Einstein simply represented the new factor by the Greek letter lambda (Λ). Instead of G=T—so beautiful an equation, so symmetrical—he would now have the hobbled G-Λ=T.

The details of how Einstein came up with the cosmological constant are subtle, but one can think of it like this: *G* represents the geometry of our universe, and it's so tightly curved that it has a high value, enough to make the stars come crashing down, like boulders falling into a vast pit. Take away a certain amount from that pull, and the stars won't crash, but will instead remain floating, fairly still, as almost all astronomers of the time believed was the case. It would be as if Einstein redrew the depth of that pit so it wasn't so deep, and the boulders no longer started tumbling headlong into it. That's what inserting the lambda did.

He was uncomfortable with the change from the very beginning. "That term," Einstein declared about the lambda from the podium in Berlin, "is necessary only for the purpose of making possible a near-static distribution of matter, as required by the fact of the small velocities of the stars." Astronomers had assured him that all the stars we saw only moved fairly slowly or randomly among one another, and this "near-static distribution of matter" would not result from his original equation. Only with the change he now unhappily put in could he stay true to what observational evidence seemed to show.

The lambda may have been necessary to bring Einstein's equation into line with the latest astronomical findings, but he felt the addition was "gravely detrimental to the formal beauty of the theory." To Einstein, simplicity and beauty were our best signs of an underlying truth. He didn't believe that any deity or force of nature would have started creating a universe in accord with ultrasimple principles, then awkwardly thrown in such a correction. The original G=T from 1915 could have been a vision of God's hand, revealing a creation that delighted in simplicity. Its two symbols arose from the nature of the universe: the G from the essence of how space curved, and the *T* from the sheer existence of things in space. The new, ungainly Λ, however, was just an arbitrary component, added to the left-hand side to make the pull of gravity weaker—in our image above, to make the "pit" of our universe less steep, so that the stars (the "boulders" in the image) would not tumble down into it.

In the string quartets that Einstein loved playing, every note had its place, every instrument its role. No one would suddenly drag a large tuba into the room and randomly blast out noise to halt the score's natural direction. That's what changing the direct G=T to the ungainly G-Λ=T was like.

But the word of the world's astronomers was unequivocal. Our sun exists in an island of stars called the Milky Way. They insisted it was not expanding, that there was just infinite blackness beyond. If Einstein hadn't so deeply believed in the need to respond to experimental evidence, he might not have put in this correction. But at that stage in his life, facts were absolutely as important to him as the sheer play of intuition. Since his 1915 equation predicted the opposite of what the facts seemed to show, then that equation had to be wrong.

This was his first great mistake.

The full effect of his error would not become clear until years later, but in the meantime Einstein tried to convince himself that his original theory wasn't a total failure. The effect that he needed the lambda to counterbalance only became noticeable over immensely large distances. Its value could be set so small that on the scale of our solar system, calculations would still be accurate, as if the original, simple equation G=T was all that applied. That's why the predictions Eddington was working with remained valid.

While he could take comfort in Eddington's findings, Einstein could not make peace with the fact that his beautiful, original theory appeared to have been fundamentally incorrect. What especially tormented him was the question of why the universe had been built to have that extra term in there at all.

Despite these inner doubts, he started to defend the ungainly G-Λ=T, accepting that the vision he had briefly glimpsed of the perfect, ultrasimple G=T was somehow not how the universe worked. He didn't love the change but became used to it.

Although Eddington's 1919 results had brought Einstein great fame and made him seem an image of perfection, the reality of his life was different. The world thought Einstein was a kindhearted, humble man at ease with how his life had worked out. Yet his second marriage was far from what he'd hoped for, and the sons he loved were slipping away.

The world also thought he had created equations of remarkable insight, approaching the wisdom of God himself. Yet Einstein, with his insertion of the lambda, knew that was a lie: either he hadn't yet reached the deepest level of truth, or the universe lacked the simplicity he so wanted to believe was there.

Part IV

RECKONING

Einstein on his favorite sailboat in Germany, 1920s

TWELVE

Rising Tensions

EINSTEIN WAS NOT alone in doubting the need for the lambda in his gravitational equation. So, too, did a Russian mathematician named Alexander Friedmann.

A veteran of the Great War, Friedmann was a mournful man whose appearance—he had a drooping mustache, small round glasses, and an expression that seemed to say he expected things to go wrong—matched his depressive nature. Late in 1914, a few months after the war had started, Friedmann wrote to his favorite professor, Vladimir Steklov, at St. Petersburg University, "My life is fairly even, except such accidents as the explosion of an Austrian bomb within half a foot, and falling down on my face and head. But one gets used to all this." Friedmann had decided to train as a pilot, which, in an atypical burst of optimism, he was doing because he had been assured "it is no longer dangerous." Steklov wrote back that this was an exceptionally bad idea.

There's a gap in the existing correspondence, but soon Friedmann was thanking the professor's wife for the warm clothing she'd sent him, which he found exceptionally useful, given his regular flights high in the frozen winter air. The professor's counsel had been ignored. He also thanked Steklov for sending him some interesting differential equations to look at, although he apologized for his lack of rigor in the solutions he quickly sent back, noting that it was hard to carry out proper investigations in his circumstances. He did, however, describe to Steklov the calculations he came up with to find the best release positions for the bombs he ended up dropping on the vast enemy fortress at Przemyśl with an accuracy that impressed, if also disturbed, its Austrian and German occupants.

Friedmann noted, too, that he was being ordered up for dogfights against the German air fleet, which, he said, was part of an army "with excellent organization and equipment," in contrast to "the lack of either in ours." Once, a German plane unleashed its new fast-firing machine guns at Friedmann. His only defense was an elderly carbine, which had to be lifted up and held at arm's length before, with a mighty bang, it would release a single bullet. ("The distance between our airplanes being extremely small . . . it gives you a terrible feeling," he wrote.) When his missions were done, he received the St. George Cross for bravery.

Surviving the war, as well as the revolution, counterrevolution, and counter-counterrevolution in Russia—not to mention poverty, lack of food and fuel, and epidemics—he came across Einstein's papers around 1920. By then, Friedmann was teaching at the Institute of Railway Engineering, as well as part time at the Geophysical Observatory in the city that had recently been St. Petersburg and now was Petrograd. Very quickly he saw what he was convinced was a flaw in the relativity papers. But how could he, in desolate Russia, convince the great German professor of what he suspected?

Back in 1917, when Einstein realized that his G=T equation might predict that the universe was changing size, he had inserted the lambda term (Λ). What Friedmann found in 1922 was that Einstein's original equation—the raw G=T, with nothing added—contained thousands, indeed millions, of scenarios for intriguing universes.

He began to explore them.

• • •

FROM EINSTEIN'S ORIGINAL G=T equation, Friedmann came up with a startling array of possibilities for how space and the "things" in it might change over time. In some of the scenarios he unveiled, a universe would steadily grow, like a sphere that inflated forever. Yet there were also scenarios—all contained in the mathematics of the original equation—in which the universe only pumped up in volume to a finite size before it then started collapsing, as if its substance was hissing out through some escape valve. Everything that mankind or other intelligent beings in such a universe had created would be annihilated.

There were yet other scenarios, in which a universe's crash wasn't final after all. Instead, after collapsing down to a single point, it would then begin to rebound back out. Everything that civilizations had built before would be utterly crushed, but the raw material would still be there to start again. Friedmann did some rough calculations: these "pulsations," he found, might recur over periods of about ten billion years.

This wasn't the first time humans had imagined such a sequence of death and rebirth. As Friedmann wrote, it "brings to mind what Hindu mythology has to say about cycles of existence," a reference to the belief that the universe has already been created, destroyed, and re-created many times. He added that, of course, his solutions could be seen only as conjecture and were not yet supported by the known facts of astronomical experience.

With his friends' support, he wrote up his findings in a brief paper, and after the best linguist in his group had improved his German—which was about on a par with Einstein's French—he daringly sent the paper to the most prestigious physics journal in the world, *Zeitschrift für Physik*. The journal quickly accepted it, in 1922. He assumed that Einstein would love the paper, for he was showing that the original 1915 equation—the simple G=T, without the random brake of the lambda—contained these extraordinary results. And if he did, Einstein would finally be able to get rid of the new term.

To the shock of Friedmann and his friends, when they managed to get hold of the next issues of the *Zeitschrift für Physik* later that year—not easy in postrevolutionary Russia—they saw that Einstein had sent in a rebuttal! The Russian's findings were unacceptable, Einstein wrote. Nor was this mere bias. Einstein had gone through Friedmann's calculations and found a flaw. "The results . . . contained in [Friedmann's] work," Einstein's published letter stated, "appear to me suspicious. In reality it turns out that the solution given in it does not satisfy [my] equations."

Friedmann was distraught. A comment like that was death to any hopes he had for further academic advancement. How could the great man do this to him? It would be too presumptuous to write a letter of complaint to the journal. Instead, Friedmann and his friends decided it would be more tactful to write to Einstein at his Berlin address. And that, again no doubt getting help for his mediocre German, Friedmann laboriously did.

Friedmann's letter to Einstein was polite but clear: "Allow me to present to you the calculations I have made . . . Should you find the calculations presented in my letter correct, please be so kind as to inform the editors of the *Zeitschrift für Physik* about it. [P]erhaps in this case you will publish a correction to your statement."

There was no response—but not for the reason Friedmann would have feared.

Earlier in 1922, when the Jewish foreign minister of Germany, Walter Rathenau, was assassinated—to the glee of conservatives across the country—Einstein realized that serious danger was beginning for prominent Jews. Already a Working Party of German Scientists for the Preservation of a Pure Science had been formed to fight Einstein's ideas. Their inaugural meeting had been held at the Philharmonic Hall in Berlin, with swastikas displayed in the hallway and anti-Semitic brochures on sale in the lobby. A few of the Einstein haters had some academic affiliation, but most were poorly educated. "Science, once our greatest pride, is today being taught by Hebrews!" the housepainter and failed art student Adolf Hitler complained.

Alexander Friedmann, early 1920s. "Allow me to present to you the calculations I have made," he wrote to Einstein, not knowing where his overture would lead.

To give the situation time to cool, Einstein took up a long-standing invitation to undertake a lengthy tour by steamship. By the time Friedmann's letter arrived, he had already left Marseille for Japan, where he wrote to his sons, "Of all the people I have met, I like the Japanese most . . . They are modest, intelligent, considerate, and have a feel for art." Friedmann's letter wasn't forwarded to him. Yet even when Einstein got back to Berlin the next year, he didn't respond.

Einstein's failure to reply was partly attributable to the vast correspondence he had begun to receive after being awarded the Nobel Prize. The letters arrived in volumes that made his previous nightmare of the roaring postman seem tame. But something else was going on — something that only the mix of fame and pride could explain.

When Einstein had first inserted the lambda term into his G=T equation back in 1917, he believed he was doing something wrong.

The Creator could not have first made the universe to be close to absolute simplicity—to have two mathematical terms, *G* and *T*, so simply explain everything about the overall structure of the universe—and then reveal it to be so different that only the addition of an arbitrary constant could make the laws of creation work.

But despite his forebodings, Einstein had reworked his equation, and now he was stuck. His reputation was at stake, for now all physicists knew his equation in that modified form. His pride was in play, too. He had done this to himself, after great soul-searching. He couldn't easily admit that he'd been weak—and wrong.

That's why he had so quickly skimmed Friedmann's paper hunting for some flaw. Then once he'd found a flaw—or thought he'd found one—he'd shown no interest in returning to the topic.

In May 1923, however, one of Friedmann's colleagues, Yuri Krutkov, managed to track Einstein down in the Netherlands, through a colleague of Einstein's who had once taught in Russia. Krutkov confronted him, politely but insistently, and proudly recounted to his sister what happened next. On Monday, May 7, he said, he was reading Friedmann's paper in the *Zeitschrift für Physik* along with Einstein. And then, on May 18, "at 5 o'clock . . . I defeated Einstein in the argument about Friedmann. Petrograd's honor is saved!"

Einstein had the decency to look back through the Russian's work and admit he'd overreacted: Friedmann hadn't, in fact, made any mathematical mistakes. He wrote to the journal editors to set matters right: "In my previous note I criticized [Friedmann's work 'On the Curvature of Space']. However, my criticism . . . was based on an error in my calculations."

The retraction was impressive, if curt. Nonetheless, back in Russia Friedmann knew he had to get Einstein properly on his side if his new scenarios were ever to be taken seriously. But how? The only way would be to give Einstein more proof. There wasn't any astronomical evidence to prove his assertions yet, but perhaps there was another way.

• • •

AS FRIEDMANN RACKED his brain to think of how to convince Einstein that the addition of the lambda had been unnecessary, the Russian used a method of imaginative problem solving that would have looked familiar to his German colleague. Specifically, he went back to the way that tiny beings who lived on a flat surface—our Flatland creatures again—couldn't step back and see their whole world. But they could do various calculations on that world, or take journeys there, which could give them the information they needed.

Friedmann imagined what would happen if one of the research centers in this flat world sent out a traveler to check what their universe was really like. He imagined such a traveler as being like a little postage stamp. If the traveler kept to a straight line and proceeded in one direction, Friedmann wrote, it would be able to look at the landscape it passed. Clearly that would alter as the traveler moved along. He would see other landscapes, other cities. But then his surroundings would begin to look more and more familiar, and finally he would find himself back at his hometown—but arriving from the side opposite to the one he had started from!

Friedmann noted: "On returning to the starting point, the traveler would find, through observations, that the point which he reached coincided completely with the point from which he had started." That's how he would be able to prove that the sphere—the "universe"—he lived on was actually finite. If, however, the traveler never found the cities becoming familiar again, he would know that his world didn't bend back on itself. This would be proof that his universe was not a sphere.

Just as with Flatland, and with our imagined Finnish skaters, Friedmann was suggesting an analogy for our own, much larger, three-dimensional universe. If we could send out emissaries to make measurements of the universe—using advanced exploratory vessels sometime in the future or just telescopes today—we could use those measurements to work out what the underlying shape of the universe was like. That would help determine which of the scenarios Friedmann had found locked within Einstein's simple G=T equation described our world and which did not. Although we couldn't actually make the lengthy journey Friedmann imagined, if our universe was truly flat, then enormous rectangles measured out in the solar system would have four right angles inside. If it were curved like a sphere—in a way we couldn't see with our naked eye, of course, or even imagine with our limited brains—then the giant rectangles that were measured out wouldn't be flat like that, but would show inner angles that splayed ever so slightly wider than 90 degrees. As the curvature went up or down, so these angles would change as well.

Friedmann knew himself to be physically feeble, and surviving in early 1920s Russia had done little to overturn his habitual depression. But yet, he had somehow survived bombing raids over Austrian fortresses and dogfights with the German air force. He had mental strength, and he also believed that he and Einstein shared a vision. The German physicist had spoken of small-scale local measurements to understand large spaces, after all. Perhaps if Friedmann managed to traverse the continent of Europe and see the great man in person, they could—together—go further.

And so, in the summer of 1923, Friedmann decided that he would imitate the miniature traveler he had imagined and journey alone to Berlin. If he met Professor Einstein in person, perhaps he could get Einstein to trust the original equation from 1915.

The year 1923 wasn't quite as bad a time to travel as when Freundlich had led his astronomical expedition to the Crimea just as World War I broke out, but it wasn't much better. Inflation had already begun in Weimar Germany. "There is a wild currency orgy," Friedmann wrote home. In the course of less than a week, a dollar could jump from being worth one million marks to being worth four million. There were poverty and food shortages, albeit not the level of shortages that Russians were living with. Even the German landscape seemed to show how far Friedmann was from home, and he was especially disoriented to see how neatly arranged German forests were, with all the trees that made them up seeming to have been planted in straight lines. There's a photograph of Friedmann from this time: the hangdog expression and drooping mustache as always, yet sporting his best double-breasted jacket and an odd beretlike cap balanced on his head; holding a sprawling mess of papers under his left arm; awkwardly gripping his shirt with his right hand, Napoleon-style, as if he doesn't quite know what to do with it; trying to smile.

He actually made it to Berlin, and even to Einstein's street . . . but then: "August 19th: My trip is not going well—Einstein . . . has left Berlin on vacation. I don't think I will be able to see him." Two weeks later, he wrote his friends again, saying that he still hoped to see Einstein. But it was not to be. At least, though, near the end of his trip to Germany, just before returning to Russia, Friedmann did visit another man who understood the disappointments life can bring. For on September 13, 1923, Friedmann made his way to the Potsdam Observatory, where he met Freundlich. The men got on, sharing their thoughts about how the universe was structured. "Everybody was much impressed by my struggle with Einstein, and my eventual victory. It is pleasant for me."

Einstein hadn't been far away, probably at his country home outside Berlin. But even if Freundlich had told him that Friedmann was around, he probably wouldn't have made the trip back to the city. He still had too much invested in the "fix" he'd made in 1917. He'd now almost convinced himself, in fact. After all, how could any Creator—or even just the rules of physics—set up a universe that would so wildly break from equilibrium? For if Friedmann was right and Einstein's original equation did show that the universe was expanding, eventually there would just be a vast aloneness of burnt-out stars and lifeless planets steadily moving farther and farther apart. That was too awful to imagine: all mankind's strivings ending up so lost. On the other hand, if one of Friedmann's other scenarios held and Einstein's original equation showed our universe was collapsing, then sometime in the future, the night sky would shine with a terrifying brightness as all the stars above came tumbling toward us. That also was too unpleasant to believe.

In the original typed draft of the retraction he'd prepared for the *Zeitschrift für Physik,* Einstein had written that despite Friedmann's mathematical correctness, the vast range of solutions described were ones "to which a physical significance can hardly be ascribed." He then thought better of that phrase and scratched it out. But he wanted Friedmann to be wrong.

The confusion was exhausting for Einstein. It would be wonderful to find clear evidence that could release him once and for all and determine if the warpings in space that his original equation predicted would do what Friedmann proclaimed or not. But that would require measuring locations in the most distant reaches of space and seeing if the stars there were speeding away, or standing still, or falling toward us. Measuring objects so far away seemed impossible. Stars might be enormous furnaces, but at their great distances they appear on earth as just tiny pinpricks of light, without observable movement of any kind.

If only someone could find a way to identify what the stars, so remote from the earth, were really doing.

INTERLUDE 3

Candles in the Sky

IN HIS JOURNAL, the Italian explorer Antonio Pigafetta recalled the moment he decided to strike out for the unknown:

"Finding myself in Spain in the year of the Nativity of our Lord, one thousand five hundred and nineteen, at the court of the most serene king . . . I deliberated . . . to go and see with my eyes a part of the very great and awful things of the ocean."

Pigafetta's decision led to his sailing with Ferdinand Magellan in 1519, in a flotilla of ships intended to reach the Spice Islands of East Asia by an unprecedented route: heading west across the Atlantic, then finding a way around or through the American continent into a new ocean that was imagined to be there. If all went well, they would circle the earth—the first time any humans had done so.

In one sense, the expedition was a success, for Pigafetta did make it back to Spain nearly three years after setting out. But he was one of only 18 survivors from the original crew of 240, which was far from the goal of bringing everyone back with which the expedition and its backers had begun.

At least the voyage started well. Magellan's crew saw marvels along the South American coast: new sorts of humans never imagined in Europe; fish that could leap out of the water ("they fly further than a cross bow–shot") yet were tracked by predators that followed their shadows and then seized and ate them when they splashed down, "which is a thing marvelous and agreeable to see," Pigafetta wrote.

But then the storms began, ferocious ones, and it was ages before Pigafetta could make the journal entry they'd all longed for: "Wednesday, the twenty-eighth of November, 1520, we came forth out of the said strait [at the southern tip of South America], and entered into the Pacific sea."

The prospects seemed excellent at first, for they encountered a great expanse of calm water. But the calmness went on, and on, with no landfall in sight—"We went fully four thousand leagues in an open sea"—and the men began to starve. "We ate old biscuit reduced to powder, and full of grubs," Pigafetta wrote. ". . . We also ate the ox hides which were under the sails; also the sawdust of wood, and rats."

Under normal circumstances, the mariners would have found their way using the stars, but in the southern hemisphere there were few of the familiar constellations to navigate by, and certainly no guiding North Star. Looking up at the strange night sky, the sailors found that "there are to be seen . . . two clouds a little separated from one another, and a little dimmed." These glowing clouds produced light "of reasonable bygnesse."

Was it a gift from God? Whatever the cause, these two mysterious glowing clouds held to the same relative position, night after night, eventually allowing the survivors to navigate their way home. Magellan himself didn't make it—speared by natives in the surf off the Philippines—but these glowing beacons in the night sky were subsequently named in his honor, becoming known as the Large Magellanic Cloud and the Small Magellanic Cloud.

Four hundred years later, Einstein would use those same Clouds to solve the conundrum of whether or not to go back to his original equation, as the Russian mathematician Alexander Friedmann had recommended. But that would not happen before the Magellanic Clouds were explored, and some of their mysteries teased out, by a second—if very different—sort of pioneer.

IN THE 1890s, in an upstairs room of Harvard University's respected observatory, banks of computers were used to analyze large glass photographic plates of the night sky. These computers were not electronic devices; rather, the word referred to the rows of young women seated at wooden tables on the observatory's second floor. Their jobs were to measure details on the plates and tabulate neatly what they found.

The observatory's director, Edward Pickering, was proud of these human computers, whom he regarded in nearly mechanical terms: "A great savings may be effectuated by employing unskilled and therefore inexpensive labor, of course under careful supervision." Just to be sure there was no dissension, he insisted that these women—some of the first female university graduates in America—not be trained in any mathematics that could tempt them to do the work of male astronomers. He also paid them very little, just twenty-five cents per hour, at a time when cotton mill workers got fifteen cents. His colleagues patronizingly came to describe the complexity of astronomical jobs in terms of "girl-hours" or, if the work was going to involve a great deal of tabulation, "kilo-girl-hours."

It takes two people, however, to make you feel bad about yourself: the one who slurs you, and you if you accept it. Few of the women accepted the men's views of them, as a ditty they composed demonstrates. To the tune of "We Sail the Ocean Blue" from Gilbert and Sullivan's *H.M.S. Pinafore*:

We WORK from morn till night
Computing is our DU-tee
We're FAITH-full and polite
And our record book's a BEAU-TY!

The most indomitable of all the computers under Pickering's supervision was Henrietta Swan Leavitt. No manipulative manager would keep her down.

Pickering's women weren't supposed to be too educated, but Leavitt had attended the music conservatory at Oberlin College and had also obtained an A in calculus and analytic geometry at Radcliffe (then called the Society for the Collegiate Instruction of Women). She was perfectly capable of carrying out the dull tabulations that Pickering assigned her. Yet she was far from content with her station, and her curiosity would get her into trouble with Pickering—and eventually change the course of Einstein's life.

Leavitt experienced a special thrill whenever the carefully packed crates from distant Arequipa in the Peruvian Andes arrived at the Harvard observatory. That's where the university had installed its great 24-inch photographic telescope, the most powerful instrument of its class in the world.

At first Pickering had sent his own brother to Arequipa to run the telescope, but after he started mailing back reports about giant rivers and lakes on Mars —which no one else looking through the telescopes could see—Pickering replaced him with another male colleague. The region was dangerous—there were, as visiting Americans noted with the typical imperialist views of the time, half-breeds in Arequipa, as well as savages in the not-too-distant Amazon— and the 8,000-foot altitude was exhausting. Also, the work was complex. The idea that a woman could ever go to Arequipa, let alone operate the telescope, was never considered.

Back in Boston, however, Leavitt had noticed something curious about the plates being sent from Arequipa, especially the ones that revealed details of the glowing Clouds that had guided Pigafetta and Magellan on their voyage. We're used to our sun shining pretty evenly, at just about the same intensity day after day. But that's because the layers of fuel in the sun burn fairly evenly. In some very different stars, the burning is highly uneven. Like a boiling pot, pressure builds up deep within and makes the "lid"—the surface

Henrietta Leavitt, probably 1890s

layer of the star, composed of shattered atoms—pop outward in long bursts of extra-brightness. That in a sense releases the pressure, so that the surface layer settles back down, and it then takes several hours or even days for the temperature to build back up and another burst of brightness to appear again.

In the smaller of the two Magellanic Clouds, Leavitt saw there were a great number of stars that burned like this. She found them by comparing plates that were taken days or weeks apart. Instead of shining with a steady fire like our sun, these distinctive stars glowed brightly at one time, then dimmed down before lighting up to shine brightly again a few days or weeks later. Since such throbbing stars had been originally identified in the constellation Cepheus, even before Leavitt began her work, they had been known as Cepheid variables.

If it had turned out that these Cepheid stars oscillated randomly, Leavitt would have found only an unimportant curio in outer space. But she began to think about it. Whenever she sent out requests for more photographs of the Small Magellanic Cloud—forwarded by Pickering, of course, for he allowed no one else to contact the on-site director in the Andes—the photos that came back always showed there was a rich density of stars, more and more with every magnification. She speculated that the Cloud wasn't anywhere near earth, but a cluster of stars at an immense distance from us.

How far away was that cluster? Before Henrietta Leavitt, no one had been able to discover a yardstick with which to measure the farthest reaches of the universe. The problem is understandable if we think about how hard it is to tell much about a single flashlight that briefly glows "on" in a pitch-black pasture at night. A medium glow could be from a strong flashlight that's far away—but it could, just as well, be from a weak flashlight that's much closer. Leavitt's great discovery was that it was actually possible to overcome this challenge when observing stars.

What Leavitt found, bent over her plates in the brick building outside Boston, was that she could sort the Cepheid variable stars like different types of flashlights. Imagine that the Small Magellanic Cloud was extremely far from us, like a very distant meadow. The Cepheid stars there would be like separate flashlights held by people standing scattered within that meadow. From our vantage point, they all could be considered approximately the same distance away.

Leavitt noticed that some of the Cepheids pulsated slowly, over a ten-day schedule. Others pulsated more quickly, over a three-day schedule. Most important, the ones that pulsated slowly were a lot brighter in the photos from Arequipa. Since she was assuming that all of them were about the same distance from the earth, this meant the ones that pulsated slowly had to be pouring out more light than the ones that pulsated more quickly. As for our flashlights in the distant meadow, if the ones that flicked on and off more slowly looked brighter than the others, we could assume they really were brighter.

By itself, that wouldn't be enough to let us find the actual distance to the meadow. But suppose we managed to get our hands on one of those flashlights—say, a slowly pulsating one—and found that it poured out two watts of light. Now when we looked at the distant field at night and saw a flashlight pulsating just as slowly, we would know that its intrinsic power was also two watts. Depending on how dim it looked at that distance, we could estimate how far away it was.

So it was with the Cepheid stars. And luckily, astronomers were able to measure one Cepheid that was much closer to the earth, at a known distance, and register how much light it actually was pouring out. That allowed Leavitt to work out a scale on her yardstick. If the newly discovered Cepheid pulsated on, say, a seven-day cycle, the Cepheids that pulsated on the same cycle in the far-distant Magellanic Clouds had to have the same intrinsic power. Depending on how dim that distant Cepheid looked compared to the near one, she could work out how far the Cloud actually was from the earth.

It was wonderful for Leavitt to be able to do this, and when one star that she was working on seemed particularly unclear, she joked to a colleague, "We shall never understand it until we find a way to send up a net and *fetch that thing down!*" Yet Leavitt also knew that she wasn't supposed to be doing this research. As one of her fellow computers wrote in a private note, "If we could only go on and on with original work, looking to new stars, studying their peculiarities and changes, life would be a most beautiful dream. But we have to put all that is most interesting aside."

Leavitt was skilled at finding ways around these obstacles, however. Once, she explained to Pickering that she had to be away from Massachusetts for a while at her father's farm in Wisconsin, but that she would very much appreciate it if he could send her personal notebooks—all of them—so that she could continue to help. What she actually worked on, of course, he did not have to know.

In 1906—when Einstein was still happily married to Marić and still trying to find a way out of the Patent Office—Leavitt put together her main findings in a paper titled "1,777 Variables in the Magellanic Clouds." She explained how peering into the Magellanic Clouds had allowed her to create a yardstick with which to measure the universe—how her Cepheids oscillated on regular schedules, and how those schedules corresponded with their actual brightness.

It was a magnificent achievement, and Pickering was furious. Leavitt was an underling, a computer, a mere *woman*. He tried putting her findings partially under his own name on papers or at conferences, but word was getting out. A Princeton astronomer, impressed, noted, "What a variable-star 'fiend' Miss Leavitt is. One can't keep up with the roll of [her] new discoveries."

Pickering couldn't bear it, and so he pulled Leavitt away from her work, explaining that she was to forget, entirely, about working on these so-called variable stars in the Magellanic Clouds. There was a thorough numbering of stellar coordinates near the North Star that he wanted her to start tabulating instead. It was work, admittedly, that other astronomers didn't consider especially important, but Pickering was a punctilious man, and with these listings he felt that he could make his name.

Leavitt repeatedly tried to get back to what she loved, and in 1912—the year Einstein was starting his collaboration with Grossmann on the mathematics for his theory of gravitation—she managed to publish a paper that gave even more details about how to use her Cepheid variables to measure true distances in the outer universe. After that insubordination, Pickering cracked down on Leavitt even more harshly. No more of the fresh plates coming from the Andes were to go to her, he decreed—not if they involved those damned Magellanic Clouds.

Leavitt died in 1921 and never did get to travel to the observatory of which she'd dreamed. A year later, however, one of her colleagues among the computers made the trip for her. Pickering was no longer the director in Boston, and regulations had been slightly eased.

Leavitt's friend traveled by steamship to South America, took trains and horse-drawn wagons to continue inland, and finally reached the top of the valley that led to Arequipa. "In the distance," a contemporary wrote, the city built of soft white volcanic stone appeared "to be a city of marble." The colossal volcanic cone of El Misti was visible to the northeast, jutting nearly four miles into the sky; Pichu-Pichu could be seen to the east. The air was thin, but the woman had to go farther, for the observatory was high above the city. When she reached it, she was more than a mile and a half above sea level, high in the crystal-clear air of the Andes.

The sun went down. The cool night began, and the stars—the brilliant, perfectly clear stars—began to appear. Afterward, Leavitt's friend took out her journal and wrote, "Magellanic Cloud (Great) so bright. It always makes me think of poor Henrietta. How she loved the 'Clouds.'"

THIRTEEN

The Queen of Hearts Is Black

EINSTEIN WAS IN A HAZE of confusion after the fall of 1923. He'd been thrown by Friedmann's unexpected paper suggesting that the original ideas in the raw G=T equation were right and the curvature of the entire universe could be changing. Clusters of stars and planets might end up sliding away from one another in what would become an infinite expansion. Or the opposite might occur, and the curvature might be flexing differently so that the ancient Hindu mythologies might prove to be true after all, and the entire universe was doomed to an endless cycle of contraction and expansion, as if we were somehow locked within an invisible sphere that deflated and inflated forever.

Einstein had managed to push aside some of this haze, at least from his conscious mind, by pretending that what Friedmann had found was merely a mathematical possibility, of no real physical significance. But then, four years after Friedmann's abortive visit to Berlin, and five years after Henrietta Swan Leavitt's coworker made it to the mountains of Arequipa, that temporary reprieve ended.

In 1927 Einstein was at a sequel to the Brussels conference he had first attended as a young man living in Prague. He was a hero now and had set aside any lingering concerns about his gravitational

Einstein and Lemaître, around 1930

equation—or at least had tried to—so as to focus on other undertakings. Yet on one of the first days of the conference, an earnest, heavyset Belgian man in his thirties came up to him and said that he had a mathematical proof that the universe was expanding.

Physics professors, even those below Einstein's level, frequently are bothered by cranks, and for Einstein this sort of thing happened all the time. He'd become good at polite but firm, immediate dismissals, and he needed that now in Brussels, where his focus was on new fields of study. But this man could not be so immediately sidestepped.

Not only was Einstein's interlocutor an official invitee to the conference—which suggested he at least had done graduate-level work in physics—but he was also wearing the stiff white collar and black woolen jacket that showed he was a Catholic priest. In fact, he was a Jesuit, part of an order that despite its dogmatic loyalty to the pope had been active in astronomy for centuries.

Einstein let the pudgy man, Father Georges Lemaître, begin to explain. He had published a paper in a Belgian journal—had the Professor perhaps heard of it?—in which he had gone through the consequences of Einstein's work, trying it with a range of values for Λ. The most interesting results arose when Λ was set to zero, so that the equation went back to its original, pure form of $G=T$.

Decades later, remembering that encounter, Lemaître said that Einstein had commented favorably on how ingenious that and other details of Lemaître's mathematical approach seemed to be. But those words were little more than the polite banalities of a famous figure trying to end a conversation, which Einstein quickly proceeded to do. Before Lemaître could finish, Einstein cut him off. Your calculations might be accurate, Einstein told him, "*mais votre physique est abominable* (but your physical insight is unacceptable)." And with that, Einstein set off to find a taxi that could take him to the lab of Auguste Piccard, the famed balloonist, whom he'd arranged to visit.

Most people would have considered that the end of the conversation. But like almost all men of his age in Europe, Lemaître had survived the Great War, in his case having served as a trench digger, machine gunner, and finally artillery officer. Events such as the world's most famous scientist walking firmly away from him and starting to close a taxi door in his face were to be considered opportunities, not rejections. The Jesuit accelerated and jumped in beside Einstein. Would the Professor care to hear how he had already taken that criticism into account?

Whether the Professor wished to or not, a moving taxi is a difficult place from which to escape. Lemaître explained that in his paper—and oh, if Einstein had subscribed to the estimable *Annales de la Société scientifique de Bruxelles,* he would surely know all this—he'd given detailed experimental evidence showing that his conclusions were true.

This was disturbing news, and suddenly Lemaître had Einstein's full attention. He had been able to push aside Friedmann by declaring that the unknown Russian's calculations were just some mathematical sleight of hand, with no astronomical facts to back them up. But now here was another scientifically trained man telling him that there was valid evidence that the universe was expanding.

Lemaître's explanation was necessarily rushed, for Piccard's laboratory wasn't too far away. He talked about the graduate work he'd recently done in the United States, at Harvard and MIT, where he'd learned remarkable things about a type of star called a Cepheid variable. He didn't know who had done the first work on those stars, he explained, but they had the ability to grow and shrink in brightness, and thus provide definite information about what was happening in distant space. That research seemed to show—the evidence was fragmentary, but ah, the Professor should recognize how significant this might be—that distant star clusters were speeding away.

Einstein wasn't rude, but Lemaître sensed that he was distracted. "He didn't seem at all well informed about astronomical facts," the Belgian remembered later. The taxi stopped; Einstein got out. Lemaître had no idea whether his message had gotten through.

It had, and it hadn't. Five years before, in 1922, Einstein had dismissed Friedmann's paper by saying that his work was just mathematics. Now, in 1927, when Lemaître went further and said he had data to back up the vision of an expanding universe—exactly what Einstein had asked for from Friedmann—Einstein dismissed that, too, as being physically unacceptable. Einstein knew that Lemaître hadn't been entirely clear in what he'd explained, and Einstein was acting as if he didn't really want to hear more, as if the fact that the findings Lemaître spoke of were incomplete and not from the most famous astronomers meant they could be ignored.

Clearly, something else was going on—and a famous social psychology experiment at Harvard suggests what it might have been. The study's organizers had a group of students briefly shown a sequence of playing cards. The experimenters had, however, reversed the colors on the cards, so that the hearts and diamonds were black, and the spades and clubs were red.

It was a study of perception. When the cards were shown slowly, the students easily saw what was wrong. When the cards were flicked very quickly—too fast to recognize any details—the students had no idea anything was wrong and also were at ease. But when the cards were displayed at an intermediate speed—that is, the subjects could just about make out what was shown but didn't have time to fully analyze it—the results were different. Many of them felt terribly uncomfortable. They complained of being dizzy, said they were suddenly very tired, or—without knowing why—said they just wanted to get out of the room. They wanted the experiment to end.

That was the situation Einstein was in after hearing about Friedmann's work and now about Lemaître's even more detailed developments. Their ideas preyed on him. He didn't entirely understand all the details yet, but he could sense the underlying truth: something was wrong, and he wanted that feeling to end.

EINSTEIN'S DILEMMA WASN'T going to be solved simply. His aversion to confronting it was too strong, and his investment in adding the lambda to his G=T equation was too great. He would need someone with more authority than an unknown Belgian priest or a Russian mathematician to move him. And in the world's astronomical community of 1927, the man at the opposite extreme—renowned above almost all others—was the director of the famed observatory on top of California's Mount Wilson, Edwin Powell Hubble.

Hubble was a man's man who in his youth, reputedly, had been such a tough boxer that Chicago promoters had sounded him out to see if he would go up against the world heavyweight champion, the powerful Jack Johnson. Hubble had turned them down and gone on to become a combat officer, serving in some of the most desperate battles near the end of the Great War in France.

Hubble didn't like to talk about the war too much, but occasionally, late at night, he would admit to awed grad students that "the hardest thing was to see wounded men fall, but yet go forward without stopping to help them." He also told of having been knocked out by shell bursts (which would explain his injured right elbow) and even of being trapped in a swaying observation balloon: terrified, of course, yet somehow finding what some people called the courage—but which he knew as simple common sense—to keep observing the battlefield below, drawing charts of the enemy's positions.

It was the story of a remarkable life—and it was far from accurate. For one thing, while Hubble was a tall man, and fit, he had done only one term of boxing as an undergraduate at the University of Chicago—a fine academic institution, but not one especially known for the ferocity of its students. No promoter could possibly have considered a relatively inexperienced student for a bout against the world heavyweight champion.

Hubble's army experience was also not quite what he described. He had been called up, but his units had never seen combat. His military discharge record has categories for battles, medals, and wounds—and the word "none" is neatly inked after each one. The elbow injury may have been incurred while playing softball during a brief stint teaching high school in Kentucky.

The advantage of Hubble's tall tales was that having Walter Mitty dreams could be a wonderful motivation for real achievements. Hubble did study astronomy and wanted to be very good at it. He ended up director of the observatory on Mount Wilson. His predecessor had been skilled at extracting money from wealthy benefactors, including a businessman named John D. Hooker, and the rugged mountain now housed the world's most powerful telescopes, including the massive 100-inch Hooker telescope. It weighed so much that the sleek iron girders and counterweights that kept it in the proper position made the inside of its curved dome look like a set from Fritz Lang's 1927 futuristic film *Metropolis.*

Hubble was further motivated to succeed by a competitor with the disconcerting habit of seeing through his mannerisms. For despite the extreme English accent that Hubble put on—his conversation full of expressions such as "By Jove" and "*Jolly* good"—he'd actually been born on a farm in the Ozarks in Missouri. So, too, had Harlow Shapley, America's other leading astronomer. Shapley was suspicious of Hubble's affectations, and like Hubble he too wanted acclaim and success.

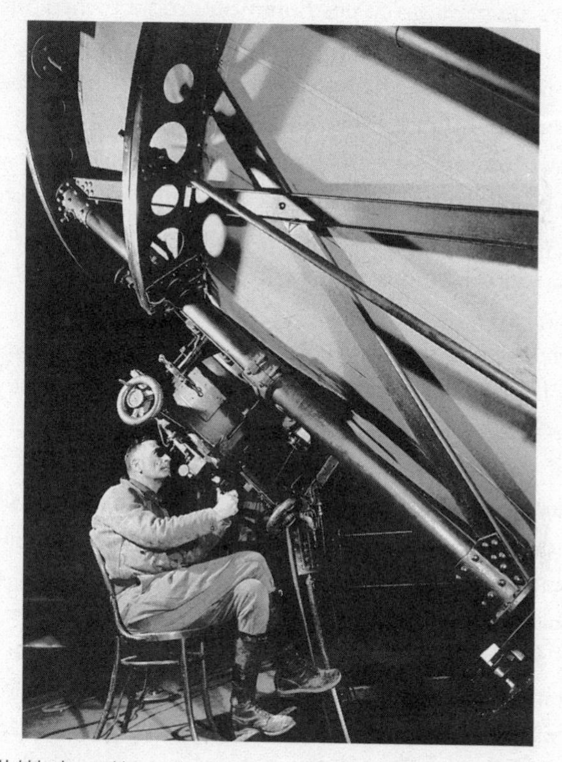

Hubble dressed in a quilted jacket for cold nighttime observations, 1937

The rivalry between Hubble and Shapley made them both keen to use their positions to spread ideas that they liked to say were their own. In 1924, for example, a Swedish mathematician had written to the Harvard observatory saying that news had reached the Continent of Professor Leavitt's stunning work on using Cepheid variables to measure distance. Could Leavitt please get back in touch to provide details on her achievement?

Normally, a communication like that from Sweden would be

taken as a sign that there was at least some interest in nominating an individual for a Nobel Prize. Shapley by then had taken over from Pickering as the director of the observatory. He wrote back to explain that, sadly, Miss Leavitt was deceased (he knew that Nobel Prizes are never given posthumously) and it turned out that he, Shapley, had actually done the main work on Cepheids, while Miss Leavitt, far from being a professor, had been little more than a passive tool under his direction.

This was the opposite of the truth, but since Shapley was now keen to broadcast news of what Leavitt had told him, her findings about Cepheid variable stars received a wider audience. This helped Hubble, still on Mount Wilson, take Cepheids the next step forward.

Astronomers at the time knew that there were a great number of stars floating nearly stationary in our Milky Way galaxy, but no one was sure whether there was anything beyond it. A few strange wisps of light called nebulae had been found that didn't fit in any obvious classification scheme, but those were generally assumed to be clouds of gas existing here and there among the Milky Way's many stars.

The 100-inch telescope on Mount Wilson was so powerful that Hubble and astronomer Milton Humason were able to take extremely detailed pictures of what was taking place in those wispy nebulae. Some of them didn't seem to be gas at all, but rather clusters of stars. The question became: how far away were they?

If these nebulae were relatively near, they'd be just more stars within our Milky Way, and the belief that the universe was composed of one unchanging island galaxy—our own—would be confirmed. If the mysterious nebula wisps were, however, much farther away, maybe we weren't as alone in the universe as we'd believed.

Hubble was a diligent worker, for as the gap between his stories and the reality of his life widened, he knew he had to achieve something substantive, and quickly, or he really would be caught out. He was good with his hands, but Humason was even better. An exceptionally careful, sensible man, he'd begun as a teenage mule driver on the mountain, helping lug materials for the observatory's construction over rough, winding trails amidst the scrubland and forests. He had trained himself, with the help of a handful of generous astronomers, to operate the heavy machinery and sensitive photographic units.

Milton Humason, around 1940

In 1925 Humason and Hubble compared several photographs of a particular nebula in the constellation Andromeda and saw that there was one star that oscillated just like one of the Cepheid variables Leavitt—or was it Shapley?—had so carefully analyzed. This particular star had a period of about thirty-one days, which was so long that Leavitt's charts showed it should be extremely bright. Yet even with the enormous magnification of the 100-inch telescope, it was very, very dim.

How could something so intrinsically bright appear dim to the observers' eyes? There was only one answer. The brilliant light this star gave out must have been diminished by its flight over enormous distances to earth. Hubble did the calculations. Astronomers often use a unit of distance called a light-year. (Despite the confusing label, this is not a measure of time, but simply how far light travels in one year. It comes to about 6 trillion miles.) Our Milky Way galaxy is perhaps 90,000 light-years across, and most astronomers at the time would have agreed that was the extent of all important matter in the universe. But the Cepheid in Andromeda worked out to be nearly 1,000,000 light-years away.

Hubble's finding could mean only one thing: our galaxy was not alone. That wisp wasn't a tiny cloud of interstellar gas or a nearby cluster of a few stars. Instead, it had to be another complete galaxy: huge, glorious, floating far away from ours, no doubt part of a flotilla in the heavens that stretched farther than anyone had imagined.

Even better than this discovery of a new galaxy was the fact that there was a way to measure how quickly it and any other distant galaxies were moving. This could be done using a variation of the well-known Doppler effect. That phenomenon initially concerned sound: if an ambulance whirs past you on the street, the sound of its siren appears to shift from a high pitch as it's approaching to a suddenly lower pitch as it speeds away. So in a sense it is also with light, although here it's not the sound that changes, but aspects of its color. A spaceship that's heading toward you will appear a little bit bluer than it would if it were stationary; if it's moving away, it will appear a little bit redder. The effect is slight at low velocities but becomes more noticeable as the spaceship speeds up.

Several astronomers had already begun to measure the shifting colors of different clusters of stars in the sky, and that's what Lemaître had used in the rough early data that he'd tried to explain to Einstein in the taxi in Brussels. The farther away groups of stars were, the redder their color became. Whatever was on the outer reaches of space really was speeding away from us.

Humason and Hubble were simply working out Lemaître's discoveries about the movement of stars in greater detail. Lemaître hadn't had such accurate information on distances as they did. No one had. Mount Wilson's great telescope was allowing them to identify pulsating Cepheids in galaxies so distant that these details would have been invisible to the telescopes Freundlich had lugged to the Crimea or Eddington had traveled with to Principe. The data being acquired at Mount Wilson also—and, oh, was this pleasing for Hubble to contemplate—could scarcely be detected by the once mighty 24-inch telescope at the Harvard research station in Arequipa that Shapley in Boston sent instructions to. Humason's telescope had a mirror 100 inches across, which could concentrate far more light than Shapley's instrument. Hubble, accordingly, couldn't resist taking a few digs at his nemesis, writing to Shapley that "in the last 5 months [I] have netted nine novae and two variables . . . Altogether, the next season should be a merry one."

By 1929 Hubble and Humason were done. Humason, easygoing as he was, didn't mind ace boxer and war hero Hubble publishing the work under just his name (though giving credit to loyal support from his "assistant" Milton Humason). The paper had a neat chart showing how far away twenty-four different galaxies were and the best evidence about how fast they were moving based on shifts in their color. There was a bit of scatter in the data, but the main thrust was clear. Galaxies were speeding away from us, and the more distant they were, the faster they sped.

The evidence presented in the paper was more complete than anyone else's, and that fact—combined with Hubble's majestic voice and skillful use of publicity—ensured that his findings traveled more widely and far more quickly than those in the *Annales de la Société scientifique de Bruxelles* could.

The news crossed the Atlantic, reaching Einstein in Berlin. And, finally, he couldn't hold out against the evidence any longer.

Einstein let it be known that lambda was now dead. Hubble had killed it—or at least he had amplified and added authority to findings showing that it was no longer needed. Einstein's original equation was restored, in all its beautiful simplicity, but his psyche would never recover.

TRAVEL WAS HARDER in the interwar period than it is today, and it wasn't until nearly two years after Hubble's 1929 finding that Einstein could make it to California, by a long steamship journey that took him first to New York and then west through the Panama Canal. There he would pay homage in person. When he and Elsa arrived in December 1930, their ship was met by thousands of excited locals, numerous photographers, and even a band that struck up a specially composed Einstein song.

If Hubble had been happy before, merely pretending that he had been a war hero and boxing champion, now in 1931, with the world's greatest scientist in attendance, his pride knew no bounds. He sent out invitations for Einstein's visit to almost everyone who counted in the American astronomical community. Elsa had already brought her husband to a great number of Hollywood dinners, for which she had an efficient if not especially polite selection procedure. As invitations poured in, she accepted them all, then at the last minute would decide which one her husband would appreciate most and cancel the others. Invitations from Hollywood royalty were always accepted, and Einstein attended the Hollywood premiere of the movie *City Lights* alongside its star, Charlie Chaplin, surrounded by photographers and crowds.

Hubble knew his own invitation was one the Einsteins were not going to cancel. On the great day itself, Thursday, January 29, 1931, Hubble dressed carefully: his shoes polished just so; his best Oxford-style plus fours (trousers bunched below the knees); his pipe; his favorite tweed jacket—perhaps a final check of the tie—and then he was ready.

The vehicle that usually trundled visitors to the top of Mount Wilson was an old, muffler-rattling truck. For Einstein's presence, Hubble hired a sleek Pierce-Arrow touring sedan instead. The photographers and newsreel cameramen who pushed close to record

Einstein with Charlie Chaplin at the *City Lights* premiere in Los Angeles, January 1931. When Einstein asked him what all the attention meant, Chaplin replied, "It means nothing."

Einstein and his wife inside the car saw beside him, on the great man's right, a beaming, exuberant, utterly content Edwin Powell Hubble.

Hubble stuck close to Einstein on the twenty minutes of hairpin turns going up the mountain, and also while they inspected the 150-foot-high tower where a solar imaging telescope was housed (only briefly—and anxiously—staying behind when Einstein proceeded to the top in the single-passenger open elevator, pulled along the fifteen-story ascent by a slender cable). Once Einstein was safely back down from the tower—and headlines of WORLD'S GREATEST GENIUS KILLED BY INCOMPETENT ASTRONOMER averted—Hubble didn't let go. He stuck to Einstein when they went into the main lodge and the other telescope buildings, and when it came time to enter the huge dome where the 100-inch giant was housed, as soon as the nimble Einstein started clambering up to the exposed catwalk at the very top—a terrifying view of Los Angeles sometimes visible in the distance, far, far below—Hubble clambered right up beside him. Photographers stood beneath them, clicking away. "He sort of wormed or muscled his way in," one colleague later recorded. "That's where he wanted to be photographed, with the great man."

After dinner, when the sun finally went down and the stars came out, Hubble escorted Einstein back to the 100-inch telescope, this time not for photo opportunities, but to look through the eyepiece and examine the planets, nebulae, and stars. Whether Hubble's greatest pleasure was in hosting Einstein or in knowing that Shapley would have to read about it in the newspapers in the following days (for Shapley's was one invitation Hubble had somehow forgotten to send) has not been recorded.

Hubble loved glory, but he wasn't selfish—at least not entirely so—and he knew it would only be fair if Humason was part of the day. When he told Einstein that this was the jolly good man who'd performed the actual recordings to get the redshifts—the data that proved how fast the galaxies were moving—Einstein settled down with Humason in one of the observatory's offices to examine the original plates. Einstein had spent years in the Patent Office in Bern and had always liked to build things. His father and uncle had, of course, been immersed in engineering throughout his youth. He respected solid, practical skills. Humason had the gnarled hands of the laborer he'd been in his youth. It was clear to Einstein, as the two men went through the images, that Humason had taken no shortcuts in his work. The shifts were unquestionable. Entire galaxies were hurtling away, at ever increasing speed.

In the observatory's library the day after the celebrated visit, in front of yet more photographers and journalists, Einstein recanted. Reading aloud, in his still not quite English-language English, he said, "New observations by Hubble and Humason . . . concerning the redshift of light in distant nebulae make the presumptions near that the general structure of the universe is not static. Theoretical investigations made by Lemaître . . . show a view that fits well into the general theory of relativity."

This was big news. "A gasp of astonishment swept through the library," the lead Associated Press reporter there wrote, for relativity mania was gripping the country. In the paper that formally announced his changed view, Einstein wrote, "It is remarkable that Hubble's new facts allow general relativity theory to seem less contrived (namely, without the Λ-term)." It was a return to the beauty he had always loved.

Einstein had accepted lambda's end as soon as Hubble's findings had come out in 1929, but his trip to Mount Wilson two years later, in 1931, giving public obeisance, made it official. *Punch* magazine in distant England was soon writing:

> *When life is full of trouble*
> *And mostly froth and bubble,*
> *I turn to Dr. Hubble,*
> *He is the man for me!*

For a plus-fours-wearing Anglophile like Hubble, that endorsement was very good. But he was also a farm boy from the Ozarks, and so a headline that appeared several weeks before that in the esteemed pages of the *Springfield (Missouri) Daily News* was even better:

YOUTH WHO LEFT OZARK MOUNTAINS TO STUDY STARS
CAUSES EINSTEIN TO CHANGE HIS MIND

FOURTEEN

Finally at Ease

WITH THE LAMBDA GONE, Einstein, finally, was at ease. "Since I introduced this term I had always a bad conscience," he explained later ". . . I [was] unable to believe that such an ugly thing should be realised in nature." The relief at being able to admit that—not least to himself—was intense.

It was too late to apologize to Friedmann, for the sorrowful Russian, undernourished, had died of typhus several years before, never knowing how much his ideas would be validated. But the portly Lemaître was still around, and Einstein was as generous as could be. At a conference in California in 1933, two years after the Mount Wilson event, Einstein stood up and said of the latest work by Lemaître, "This is the most beautiful and satisfying interpretation . . . I have listened to."

Later in 1933, back in Brussels where they'd first met in 1927, not only didn't Einstein try to slam a taxi door in the priest's face, but he announced at a conference that Father Lemaître would have "some very interesting things to tell us"—which sent Lemaître into a flurry of activity before the next sessions, for he'd not known he was going to present. When Lemaître did pull together an impromptu talk, Einstein could be heard from the floor, loudly whispering in his Swabian-accented French, *"Ah, très joli; très, très joli* (Very beautiful; very beautiful indeed)."

Einstein was happy not just because he'd been proven right in his original symmetrical view of G=T. He also now saw that Hubble's findings allowed us on earth to be like those Flatlanders in Edwin Abbott's fantasy who managed to step beyond their universe and see what was really happening. A. Square had needed a visiting sphere to help him. Friedmann had suggested—also metaphorically—sending out a traveler who would march in a perfectly straight line and see if he ended up back home. Neither of those was likely to work here, but the cartographic technique that Grossmann had taught Einstein—the simple tactic of measuring the angles of triangles and rectangles to see if the figures were flat, or if their surfaces were swollen outward—was closer to the actual solution Einstein could now use.

It was a solution that Hubble himself couldn't quite grasp. He understood that the Cepheid stars in Andromeda showed that our galaxy was just one of many other galaxies—immense islands, each containing billions of stars, that stretched far out into space: to the very limit of what the new 100-inch telescope in the mountains of California's dry desert could detect. And Humason's redshifts showed that those galaxies were moving away from us, and fast—and the farther away they were, the faster they were moving.

That was about as far as Hubble could go, for he would have been the first to admit that he was no theoretician. Einstein's work already had strange consequences: the fabric of empty space rippled when Hubble climbed a ladder and "pushed" through it, and wiggling his hand made space dip and sag around it. The latest findings were even more startling. The observation through the 100-inch telescope that the far galaxies were speeding away from us would make sense if the universe had been created on a mountaintop in California, and—as if from a cataclysmic magma burst—everything was still moving outward from that point. But even the immodest Hubble couldn't quite believe that all the distant galaxies in our universe knew where he was and that he'd be at the center of all future events, watching them recede.

The true explanation was more humbling. Imagine you're holding a deflated child's balloon, colored white. Now with a red marker pen put a dozen red dots on it. Start blowing up the balloon, and you'll see that the dots will start moving away from one another.

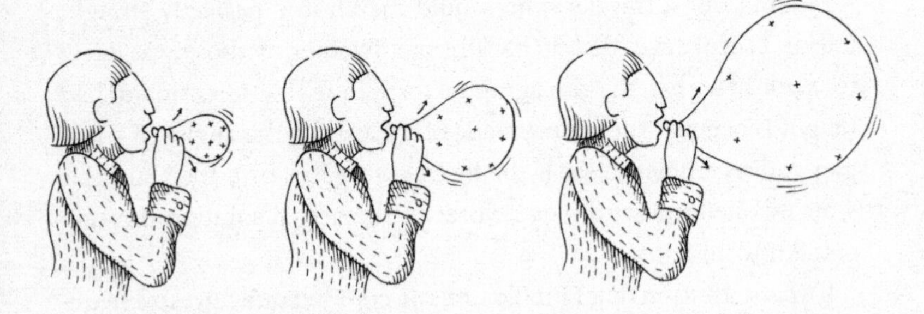

Even better, dots that are near one another will be separating slowly. Dots that are far away will be separating faster. It doesn't matter where you start. Look at one of the dots that happen to be at the top of the balloon. As you puff away, the dots nearest to it will move a small distance. Dots on the far side will move faster, with the whole bulk of your breath inside propelling them away. Now shift your attention to a particular one of those distant red dots. In the same amount of time that it takes for the ones nearest to it to only move a short distance, the ones farther from it will go much farther.

The effect becomes more dramatic if you imagine this happening on earth. You're standing in London, at the Houses of Parliament, and you see the bucolic wonderland of Battersea, on the other side of the Thames, begin to move slowly away from you. That's not too surprising, for you observe that the Thames is widening at the stately rate of 1 mph. But radio reports come in that Dublin is moving away from you at over 100 mph, and that New York—even more distant—is moving away at a rate of 3,000 mph.

That might make sense if there were some giant lava flow under the Thames pushing the earth apart, with London at the center. But then other reports start coming in, strange ones. A BBC reporter in New York insists that he feels *he's* standing still. The New Jersey shore is moving away from him at a stately 1 mph, as the Hudson River slowly widens. But Toronto, farther away from New York, is receding at 300 mph, and it's even more distant London that is receding at 3,000 mph.

This is weird, for how can both London *and* New York feel as if they're the static epicenter of some giant planetary lava flow? This can only happen if the entire volume of our planet is expanding. What happens on the surface might appear odd—those cities running away from each other in this uneven way—but if you view the earth as a giant balloon or beach ball that's inflating, it makes perfect sense. Nearby cities move apart slowly. Distant cities—distant points on the surface of the planet—move away from one another more quickly, as the entire sphere inflates.

That was effectively what Milton Humason had measured in outer space. The distant galaxies are like the dots on our balloon or the cities on our planet. And the fact that not only are they moving apart, but that regardless of which dot you are standing on, the nearby ones are moving slowly, and the farther ones are moving more quickly, can mean only one thing. What seems to us as our complete universe—the three-dimensional space we live in—is actually just the surface of something else: something huge; something terrifying. A two-dimensional balloon expands in a way we understand into three-dimensional space. By analogy, our three-dimensional universe, with all our galaxies and planets, must be expanding into four-dimensional space—a logical consequence that our limited minds can't possibly visualize.

To Einstein, Humason's finding was what he'd always hoped for.

The prediction contained within his original equation—what he'd mistakenly pushed away when Friedmann and Lemaître had tried to show him—was right. Our universe is just the surface of something like a giant sphere. Galaxies are scattered all over its surface, and at the moment they're flying apart from one another, as the "underlying" sphere is expanding. We in our Milky Way are not special; no particular galaxy is. We're all just dots on—or within—an expanding balloon. That's disorienting for us "flatlanders" to imagine—but it had to be true, given the unambiguous measurements taken on Mount Wilson.

BY THIS TIME, in 1929 and the years right after, Einstein was much more at ease in his personal life. He and Marić had achieved an understanding, in large part because Michele Besso had acted as a calming intermediary. Einstein had also felt it only fair to give Marić the substantial Nobel Prize payment he received. She invested most of the money in rental properties. That she was financially secure made her less bitter, which in turn helped Einstein become closer with his sons. After one holiday with the boys, Einstein wrote to Marić that their good behavior showed "you have proved that you know what you are doing."

Life was also improving with Elsa. When he'd first met her, he'd written, "I must love someone, otherwise it is a miserable existence. And that someone is you." That initial burst of love had faded after their marriage in 1919, but gradually a surprising amount of it came back. Even though Einstein continued to have affairs, he never humiliated her directly, he was always generous, and he had a sense of humor she loved. He also recognized that even an imperfect marriage could develop its own satisfactions. Elsa thought the world of him; she was an excellent hostess, always putting people at ease; and he enjoyed her nicely ironic sense of humor.

In December 1930, for example, when they arrived in California for the inspection of Hubble's results, among the waiting crowds were several dozen cheerleaders. The sight had struck them as so preposterous that Elsa decided to inspect them as if they were a military guard—walking along and murmuring suitable comments about their appearance, much to her husband's amusement.

Nothing flummoxed her. Another time, when visiting the University of Chicago with Einstein, she talked about a recent stay in Princeton, saying that she and her husband had liked it very much, despite the difficulty with the flying snakes. The interviewers were confused, so Elsa elaborated: the flying snakes that bit her on the hands. They were more confused, so she went on: the same flying snakes that had flown up under her skirt! It was at that point that a bilingual hostess stepped in. Was it really flying snakes? she asked Frau Einstein in German. Elsa shook her head. Americans could be so naive. *"Nein!"* she explained. *"Ich spreche von Schnaken* (I'm talking about mosquitoes)!"

At their home in Berlin, Elsa took great care to ensure her husband's comfort. Einstein loved fresh strawberries, for example, so she bought them whenever possible. The couple had a blue parakeet, which made the kitchen pleasant, and they hosted musical nights, too. Einstein also had plenty of time to relax with the piano or his beloved violin on his own, even though the neighbors did not appreciate him playing it quite so vigorously in the echoingly tiled kitchen at night.

Even stays at their summer house often led to good times. Einstein loved sharing walks and the beautiful views with Elsa and his stepdaughters. His son Hans Albert, now more reconciled with him, at least once showed up on a motorcycle, to everyone's fascination. There was mushroom hunting in the woods, the strange "yo-yo" toy a neighbor's son let them try, and the fruit trees and shady porch. It was to Hans Albert that Einstein had said his wife was "no mental brainstorm," but then he had gone on and added, "[Yet] she is exceptionally kindhearted."

Elsa's daughters seem to have taken their stepfather's side and concluded that the trade-off of accepting his affairs was more than worth living with "Father Albert." And whenever he really had to,

Einstein drew back to protect his marriage. In 1924, for instance, he had written to one exceptionally besotted young university graduate that there was going to be no future for them and that she should simply "find somebody who is ten years younger than me and loves you just as much as I do."

As Einstein's family life had stabilized, he had achieved equilibrium in other ways as well—or so, at least, he thought. The reason he felt this way can be seen in his reaction to a particular contribution by the man who had once been a thorn in his side: Lemaître.

Back in 1927, before Einstein had decided to get rid of the lambda, he had been rude to Lemaître, not giving his work serious attention. This had hurt the inexperienced Belgian, leaving him dejected. After finally getting Einstein's support, however—as well as that of Eddington and everyone else who counted—Lemaître's confidence was back. He began to look a bit more at the dynamics he'd pulled out of Einstein's raw equation. The universe might be expanding, or—in line with Friedmann's vision, which so uncannily matched Hindu myths—it might be constantly cycling back and forth, as if it were "bouncing" in size. Yet both of those views presumed that this was a process that had always been taking place: that there had been no creation, just as there would be no end.

Why?

For the rest of his life, Lemaître insisted that what he did next had nothing to do with his religious beliefs—that religion was one path to the truth and science was another, and the two could operate quite independently of each other. But papers discovered after his death show that even when he was still in training for the priesthood, in the seminary, he'd jotted a note to himself: "As Genesis suggested it, the Universe had begun by light."

Now, newly confident in the years after 1929, he began to see how that idea, too, might be hidden within Einstein's raw equation. Couldn't one simply travel backward in time and see what it all must have started from? With the Mount Wilson measurements, these sorts of reflections were not entirely theoretical anymore. Humason had shown that some galaxies were hurtling away from us so fast that yesterday they had been perhaps a billion miles closer, and the day before that two billion miles. All the galaxies beyond our local cluster had once been closer. It was as if a giant grenade had gone off long ago, sending fragments—these galaxies—flying outward. We arrived very late on the scene and could only see those flying fragments. But in our mind's eye, we could work backward, and backward, until we reached the initial moment of the explosion—what Lemaître called "A Day Without Yesterday."

Lemaître published his new calculations in 1931. They were more complicated than the preceding summary, for instead of imagining the primordial "atom" as a glob of matter inside a region of space, we had to imagine space and time itself whooshing to a more tightly compressed point. Our mathematics can be precise, but our mental visions—and our words—have to remain metaphorical. Lemaître gave it a go, saying, "The evolution of the universe can be likened to a display of fireworks that has just ended: some few wisps, ashes and smoke. Standing on a well-cooled cinder, we see the fading of the suns, and try to recall the vanished brilliance of the origins of the worlds." That was, in fact, what Einstein in 1933 called "the most beautiful and satisfying interpretation of creation I have listened to."

Lemaître's theory of the origins of the universe was stunning. It was revolutionary. And it—like so many other seminal achievements in theoretical physics—owed everything to G=T.

Both good and bad came from the rehabilitation of Einstein's original gravitation equation. The pleasing consequence was that Einstein—and all those who understood his equation—had just seen one of the most astonishing aspects of science: humans can write accurate equations that are "smarter" than the people who devised them, in the sense that these equations can produce stunning, accurate predictions that their creators never realized were there. A mere mortal, sitting in his study and strolling the streets of Zurich or Berlin, had been able to use pure thought to come up with the idea that G=T, and in so doing had opened the floodgates of the many astounding—and, frankly, unimaginable—predictions that then came pouring out of it.

Even more satisfying for Einstein was the belief that his thoughts had revealed that the universe is tidy: built on exquisitely clear principles. This architectural unity is what Einstein had always loved. In getting rid of the lambda, he received confirmation that this crisp reality truly was out there, waiting for humans to discover it.

The other consequence was less positive.

Geniuses have to push hard to come up with their first ideas. Almost always, they're going far past what everyone assumes to be true and have to be confident that they're right. That involves being stubborn. But they also need to be supple, making sure their breakthroughs incorporate all the relevant factual information, then keeping their later work responsive to what others are finding out, too. The trick is to surf this line between the supple and the stubborn without straying too far to either side.

Einstein was about to break that balance. He had only added the ungainly lambda to his equation because Freundlich and the other astronomers working in 1915 and 1916 hadn't known about the universe's expansion. If they had possessed all the facts, they never would have opposed him, and he wouldn't have done such a thing. Never again, he vowed, would he be duped in the same way; never again would he let the limited state of experimental knowledge make him undermine what he was convinced was a pure, attractive theory.

Years later he apparently told a colleague that putting in the lambda was "the greatest blunder of my life." But about this he was wrong. Einstein made an even greater mistake by then deciding that he could ignore experiments that seemed to disprove what he was convinced was right. He'd made that error in his dealings with Friedmann and Lemaître, but he had made it in other regards, too.

Over the years, Einstein had bumped up against other experimental evidence suggesting that the universe was less tidy than he believed. He had never wanted to accept that. Now his experience with the lambda had made him downright obdurate—and less inclined than ever to accept disagreeable findings about how the cosmos actually worked.

Part V

THE GREATEST MISTAKE

Einstein, early 1930s

FIFTEEN

Crushing the Upstart

IN ALL THE YEARS that Einstein had been working on large-scale questions of the universe's structure, physics had also been making advances in the realm of the ultrasmall, at the level of atoms and electrons. This was happening at the same time that Einstein had conceived of G=T, and later as he adulterated it with the lambda, and still later during the more than ten years he had uncomfortably abided that unwanted term's existence. An entirely new view was taking shape. It represented as big a jump in our understanding of the world we inhabit as that which the Victorians had created in their physics of the century before, and that which Einstein's theories of special and general relativity had effected during the twentieth century. This revolution would threaten everything Einstein held to be true, and his response would lead to the scientific isolation that he endured at Princeton.

The old paradigm had been kind to Einstein, and he had grown comfortable with it, even as other physicists were working to overturn it. When Einstein had been young, and even into his twenties and thirties while he was achieving so much with the ideas leading to G=T, thinkers had assumed that whether you looked at large objects or small, it would be possible to find precise laws that explained how they moved. Yet by that point in Einstein's life, evidence had been emerging to suggest that this was not the case—even if his fellow scientists had a hard time accepting that interpretation at first.

In 1908, for example, when working in Manchester, the bluff New Zealand–born researcher Ernest Rutherford had discovered something that seemed too strange to comprehend. He had fired tiny particles into thin sheets of atoms, and although most flew right through, or got deflected off course by a few degrees, there were a small number of particles that bounced directly back.

"It was quite the most incredible event that has ever happened to me in my life," he wrote. "It was almost as incredible as if you fired a 15-inch shell at a piece of tissue paper and it came back and hit you."

Rutherford's discovery challenged all expectations about how subatomic particles would behave—yet the rebound effect he discovered didn't end the view that everything could be understood with exact and causal certainty. After only a few weeks of confusion, Rutherford worked out that what this really meant was not that there was random chaos inside an atom, but rather that there was something very hard in there. That hard bit at the center of each atom, he realized, could be seen as being like a miniature sun. Flying around it, he imagined, there would be miniature planets. Those were the far lighter electrons. The particles he'd shot into the atoms had mostly passed through the empty space between the miniature "planets," but occasionally one had hit the tough "sun" at the center—what he came to call the atom's nucleus—and that's why they had been deflected back.

This was a comforting and familiar interpretation—the idea that the micro-world operated just like a miniature copy of the macro-world; that we humans lived on a planet in a big solar system, and inside us there were a multitude of smaller "solar systems" making up the atoms of which we were composed. None of this undermined the standard view of how science advanced: that with ever greater analysis and more powerful tools, scientists would continue to see precise activities, however deeply within matter they entered.

Then, in 1912 and 1913, the Danish scientist Niels Bohr worked out even more details about those miniature "solar systems" Rutherford had discovered. While Rutherford looked like one might imagine any stocky New Zealand farmer to look, Bohr resembled no one else. He had a great wide head and unusually big teeth. When he and his brother were toddlers, a passerby apparently once commiserated with his mother on having such clearly abnormal children. Yet he was also an exceptional soccer player. At his Ph.D. ceremony, the faculty at the University of Copenhagen were discomfited when they realized that many of the attendees were simply other soccer players out to support their outstanding teammate. Bohr's even more skilled brother was a star of the national Olympic squad, and the story goes that when Niels later won the Nobel Prize, one head-

Niels Bohr on holiday in Norway, 1933

line in a sporting paper read BROTHER OF SOCCER STAR WINS PHYSICS PRIZE.

Bohr mumbled when he spoke, which he did with unusual slowness, but he was the kindest of men, with a profound creative mind, and took delight in friends who shared his ability to take a fresh view of life. When Bohr had begun his work on electron orbits, for example, he had been studying with Ernest Rutherford and staying at a boardinghouse in Manchester. The students living there suspected that their landlady was recycling the Sunday roast, turning it into dishes so many days or weeks later that it was no longer fit to eat. One of the students was a Hungarian, George de Hevesy, who after reflection decided to spike their Sunday leftovers with a radioactive tracer from Rutherford's lab. A Geiger counter–like device that was smuggled in many days later showed that the young men's suspicions were true. Bohr and de Hevesy became lifelong friends (and indeed de Hevesy later won the Nobel Prize for his work on radioactive tracers).

In Bohr's research into the architecture of atoms, many of his early findings seemed too strange to incorporate into the march of rational physics. He realized that electrons couldn't really operate like the miniature solar systems Rutherford had imagined. If electrons did start out circling the nucleus, they would soon end up tumbling inward, and the atoms would collapse. As, however, we, and our planet, and much of the universe are made up of atoms that *haven't* collapsed—as our bodies haven't shriveled to concentrated particles of dust—something else must be going on to keep the whirring electrons more stably in position.

But this strange aspect of electrons' orbits was, like Rutherford's discovery about atomic nuclei, something that could be understood in what were still fairly conventional terms. Bohr came up with the notion that electrons were locked into a fixed range of possible orbits. They couldn't randomly glide down from a distant position to one closer to the central nucleus. Instead, they were restricted to making tiny hops from one particular orbit to another. It would be as if Neptune could suddenly appear whirling in orbit just beside the earth, or possibly beside Mars or another planet, but it could never appear anywhere else in the solar system. This concept, like Rutherford's theory, was strange to imagine—but once it was accepted, there was no inherent limit to the detail with which the underlying phenomena could be described. Those hops came to be called quantum jumps (in the sense of "quantity").The term emphasizes the way these hops occur in discrete, fixed amounts.

THE CLASSICAL VISION that Einstein had been brought up with was being stretched but still hadn't broken. In fact, he had been a central player in many of the first twentieth-century advances in the realm of the ultrasmall: so successful that his Nobel Prize wasn't for his large-scale studies such as G=T, but for work he had done in 1905 explaining how light could be a particle and a wave at the same time. The particle side could be used to explain the way metals so often send electrons shooting out when they are hit by light. To the outside world, this idea seemed another mark of his genius, but to Einstein it only made sense: the universe always has an order, which human reason can find.

A decade after his 1905 findings about photons, in his exuberance after first working out G=T in Berlin, Einstein had taken his early work on subatomic particles even further. In the summer of 1916, resting after his exhausting research that had led up to G=T, he detailed how electrons that weren't otherwise liable to plummet down from "higher" orbits around their atoms could, sometimes, be excited if we pumped in extra light to strike them. When that extra light then made those electrons "fall," they released their own blasts of light, like Lucifer plummeting down from heaven. That could lead to something of a chain reaction: producing in this case not a deadly atomic explosion, but simply pure, useful light.

Einstein wouldn't have been able to construct a machine to keep this process going with the limited equipment available in wartime Berlin. But this Light Amplification through the Stimulated Emission of Radiation—the acronym for which led to the name "laser" —would ultimately be understood by his fellow researchers. In this seemingly casual paper, Einstein had laid out the basic dynamics of the laser: the device at the heart of today's fiber-optic cables, and without which the Internet would not operate. And since he couldn't know when the jumps were made, he had introduced the probability of their occurring with no cause.

The big question was whether these ideas about photons, electrons, nuclei, and other subatomic objects still fit within the underlying certainty that all of science, since Galileo and Newton, had been finding in the world. Einstein believed that they must—yet his conviction that the universe was governed by orderly, logical principles was increasingly at odds with the latest research. For instance, Einstein disliked the way, at least in his preliminary findings, he couldn't tell exactly which electrons were going to be knocked out of their orbits first. "The weakness of the theory," he wrote in his published report, "lies . . . in the fact . . . that it leaves the duration and direction of the elementary processes to 'chance.'"

At the time, Einstein wasn't too deeply bothered by the randomness inherent in his theory about light being released from falling electrons. In many other fields, we make do with statistical averages: the heights of recruits into the French and German armies; the color of leaves in a forest at a certain time of year. None of that is taken to mean that randomness truly prevails. We feel that if we looked more closely, we'd be able to trace the sequence of events that led to each recruit being a particular height, or each leaf taking on a particular hue. The common view is that this sort of recourse to statistics, to probability, is not fundamental, but just a convenient shortcut when we're not able to examine the detailed causality behind each particular object—that if we were to look into those details, the need for probabilities would disappear.

Einstein's belief that randomness would eventually be dispelled from his theory explains why he put the word "chance" in quotation marks. He knew that within his calculations, it was helpful to talk about the probabilities of the various sorts of transitions. But at heart he remained a classical physicist. He put in the quotation marks to show his belief that if we had the time to look at the details, we'd no doubt see that each transition had simple, precise causes. "The real joke presented to us here by the eternal riddle-setter," Einstein told his friend Besso, "has absolutely not yet been understood."

Einstein had faith that the great riddles of the universe could be answered in a logical way. By the mid-1920s, however, results were coming in that seemed to violate that promised clarity—and this is what set Einstein on a collision course with his fellow physicists in the burgeoning study of the ultrasmall.

AS SUBATOMIC RESEARCH progressed into the 1920s, it became increasingly clear that this minuscule realm seemed to follow principles that were far more unexpected than anyone had imagined. Although atoms as simple as hydrogen followed the principles that Bohr had laid out, more complex ones—carbon, gold, aluminum—seemed to have electrons that acted entirely differently. Arnold Sommerfeld and others made attempts to jury-rig fixes and make everything continue to operate by conventional means, such as imagining that the electrons weren't entirely like solar system planets whirling around the central nucleus all in neat circles on one plane, but instead were following ellipses or flying in complex three-dimensional patterns around the nucleus. But all were stopgaps.

In 1924 Einstein's friend Max Born, a professor at the great German university of Göttingen, told his top graduate students and teaching assistants that he was fed up with these half measures and wanted to try to find a theory that could address them. He was nearly Einstein's age, and might have been expected to resist the surprising new phenomena that were so different from what he'd been taught. But although Born was a strong thinker, he was nowhere near Einstein's level—and that actually gave Born an advantage, for it meant he didn't have as much invested in his own past achievements as Einstein did. It was the classical approaches that had been so tremendously productive for Einstein. Born, by contrast, had less to lose in jumping to a new view.

Born and his students knew that Isaac Newton had managed to work out the mechanics of the large-scale, visible world we inhabit —of trees, and moons, and powerful steam engines. It was the job of present-day physicists, Born now insisted, to do the same for the underlying micro-world where the new, minuscule "quantum" jumps were taking place. That fresh science—if it could be created —would be called quantum mechanics.

A year later, in 1925, the brightest of Born's teaching assistants, a handsome, blond-haired, and highly strung twenty-four-year-old named Werner Heisenberg, managed to solve Born's problem. Heisenberg was a great believer in German romanticism; he loved hiking in Germany's hills with muscular young men and dreamily watching sunrises. After several months of work, his inspiration came together in an intense burst one night on the North Sea island of Heligoland, on whose clean, windswept beaches he'd sought to escape the hay fever he suffered from on the mainland.

Heisenberg succeeded by putting aside, entirely, any attempt to work out exactly how the electrons in an atom were flying about —whether they were tracing ellipses, flying high over the "north pole" of the nucleus, or following some other pattern. He knew that Einstein, his hero, had achieved great things in relativity by simply looking at what we could measure of an event—be that waking up in a falling elevator or seeing an innocuous piece of radium metal glow with pure energy—without always trying to imagine the details of why it worked that way.

Now, for his own purposes, Heisenberg made lists of what investigators could observe of the light that electrons produced under different circumstances. Those observations changed as the atoms of which the electrons were a part were bombarded with light or oth-

Werner Heisenberg, 1926, one year after his breakthrough on the windswept island of Heligoland

erwise stirred about. He was simply going to record what went in, and record what came out, and work out the simplest mathematical operations to link the two together.

As a very rough analogy for what Heisenberg was attempting, imagine taking note of the clothes a large number of actors were wearing as they hurried backstage to change between acts in one of the large operettas then so popular in Berlin. From that, one wanted to work out how that corresponded with the clothes they were wearing as they came back out for the start of the next act. Some patterns would be clear. Watching the performance, one might see that women dressed as princesses were likely to come out as peasants (if, for example, the story line was moving from a palace to the countryside). Such an analysis would be limited, but in Heisenberg's new approach, it would be enough. No one would need bother to try to see the rush of individual changes taking place backstage;

all that would be measured was what we could observe, appearing "somehow" from behind the curtains.

Heisenberg's whole process was not that different from the approach Einstein had taken in his early laser system in 1916. There, one arrangement of light photons goes in, and a different one comes out. We can measure them and get good at predicting how the former will lead to the latter. So it is with the musical theater analogy —and so it was with Heisenberg's formal calculations on Heligoland in 1925. He could tabulate a range of possible events inside an atom and from that calculate the spectral lines that were seen. As to what "actually" went on inside the atoms to create the output we saw—whether it was inherently unknowable or just too complex to understand yet—was not something that, at that point, he was going to speculate about.

Heisenberg had accomplished what none of the older physicists working on the problem had managed. With the achievement of a lifetime lying in scattered notes on his desk ("It was almost three o'clock in the morning . . . I was far too excited to sleep"), he hiked down to the southernmost tip of Heligoland, climbed a rock jutting into the sea, and—as Einstein and his friends had on the mountain near Bern twenty years before—rested there to watch the sun rise over the North Sea stretching before him. Strict causality had triumphed in the West for hundreds of years. Now, limiting himself to external measurements just as he thought Einstein had—presuming it was not our job to speculate on what went on "inside"—he had a different breakthrough. Heisenberg's work is considered the birth of the new quantum mechanics.

As soon as Heisenberg returned to the German mainland, he told everyone what he had achieved. So long as one didn't worry about tracking the final details inside an atom, he explained, remarkably accurate predictions about the light it would spray out could be made. Since the great Isaac Newton in the seventeenth century, science had been built on the assumption that, at least in principle, clarity could be found about every process we observe. Heisenberg seemed to be saying that didn't have to be true.

Max Born accepted the new approach, pretty much, because Heisenberg's results were so accurate. Einstein didn't, but he was friendly with the whole Born family, and so had to tread carefully. He wrote to Born's wife, with well-chosen ambiguity, "The Heisenberg-Born concepts leave us all breathless, and have made a deep impression."

Einstein also was ambiguous because, though he objected to the way Heisenberg seemed willing to put causality to one side, he knew that physicists often missed out on important discoveries when they were too set in their ways. In 1895, for example, the German Wilhelm Röntgen had described the strange phenomenon of X-rays, and physicists who refused to accept the finding were soon proved wrong. But judgment has to be involved. In 1903 a distinguished French physicist described the equally strange new phenomenon of what he called N-rays, yet within two years they were shown to be just an experimental flaw, and physicists who had *not* resisted were proved wrong. Einstein wasn't going to give a final public statement about Heisenberg's work yet.

The Borns, for their part, suspected that Einstein was simply being polite. When Max Born probed, Einstein explained to him more of what he believed: "Quantum mechanics is certainly imposing. But an inner voice tells me that it is not yet the real thing." To a closer friend, Einstein was even blunter: "Heisenberg has laid a big quantum egg. In Göttingen they believe in it. I don't."

Soon Born had to tell Heisenberg that Einstein wasn't persuaded —and Heisenberg couldn't bear that. His friends knew that although he tried hard to give the impression of being in control of himself, he was always on the edge of becoming frantic when he felt stressed. This was especially noticeable when he would start slamming out romantic pieces on the piano with a terrifying intensity. He loved being dominant, strong, triumphant. His insight into atoms should have been the achievement of a lifetime. Now the most respected thinker in the world was saying his insight was wrong.

Perhaps the answer, Heisenberg decided—much as George Lemaître would later do—would be to speak directly to Einstein and clear everything up in person.

SIXTEEN

Uncertainty of the Modern Age

HEISENBERG HAD NO IDEA how deeply Einstein opposed the theory he'd come up with that night on Heligoland.

To Einstein, probabilities were just a sign of gaps in our understanding. They were temporary fixes that, when science caught up, would be replaced by clearer understanding. After all, Uranus's orbit had been a mystery until nineteenth-century astronomers had worked out how the unseen planet of Neptune was tugging on it. Infections had been a mystery until microscopes and other lab techniques became sophisticated enough to identify microbes.

Einstein believed that whatever was waiting in the outside world to be discovered couldn't depend on the quirks of who the observer was or how he or she traveled. He'd had hints of that objective reality sitting contentedly with his pipe and a book at the cafés in Zurich, ignoring the busy student life around him, or sitting equally contentedly with a work pad and his pipe amidst the pandemonium of toddlers and guests in his and Marić's first apartment in Bern. It arose even in the steady bemusement with which he'd viewed his great fame after 1919. Events seemed to rush past, to confuse us, to be chaotic: languages and cultures and children and words. But that was just appearances. Studied carefully enough, they were always very exact, very certain. That's why he was proud, but not surprised, to have found the certainties of relativity.

Quantum mechanics, however, did not fit with that worldview.

There was good historical precedent for Einstein's perspective. One of his greatest heroes was the Dutch Jewish philosopher Spinoza, and although he had lived three hundred years before, Einstein took solace in the fact that Spinoza, too, "was convinced of the causal dependence of all phenomena, at a time when the success accompanying the effort to achieve [that knowledge] was still quite modest." Had Spinoza been able to live long enough, he would have seen our technological civilization find precisely the causal links that he had imagined to be there and use them to create our cities, our trains, and our aircraft.

There was an even deeper reason that Einstein was so enamored of the concept of causality. He didn't believe in the tenets of revealed religion—he didn't believe there was a divine force behind Moses's tablets on Mount Sinai; he didn't believe in the resurrection of any rabbi, however wise, from Galilee—but that is far from meaning he wasn't religious. He thought that being an atheist was presumptuous, and he was awed at the intelligence manifested in natural laws. "This feeling is the guiding principle of [a scientist's] life and work," he wrote, "in so far as he succeeds in keeping himself from the shackles of selfish desire."

So the very core of Einstein's intellectual and spiritual life depended on the premise that all underlying reality was clear, exact, understandable. He was not going to believe that the universe was fundamentally unknowable.

In our earlier metaphor of actors changing clothes at a theater, Heisenberg would have been convinced that what had happened backstage was inherently a blur. From Einstein's perspective, that was wrong. Obviously, each individual actor had to be changing his or her costume. It might be hard for us to see, peering into the poorly lit changing areas, but the fact that they all came out with different costumes proved that this had happened. The way electrons moved inside an atom, Einstein felt, was just the same.

Since Heisenberg knew little of Einstein's deeper feelings, he still felt the great man could be persuaded. In early 1926, Heisenberg was invited to give a lecture in Berlin, which he knew Einstein would attend. Afterward, they fell into discussion, and Einstein invited him home. After exchanging pleasantries—Einstein asking about Heisenberg's favorite teacher, Arnold Sommerfeld, whom he knew well—Heisenberg mentioned what was bothering him.

In the 1916 work on light hitting atoms, Heisenberg pointed out, Einstein hadn't tried to describe what was going on inside individual atoms. He'd merely described what went in and then what came out. Heisenberg explained that this was exactly what he'd been trying to do in his great nighttime breakthrough on the island of Heligoland. Even so, Heisenberg remembered later, "to my astonishment, Einstein was not at all satisfied with the argument."

"Perhaps I did use such philosophy earlier," Einstein answered him, ". . . but it is nonsense all the same." The question of what was observable in relativity was very different from what was observable in the micro-world, he explained. What had taken place in 1916 was just preliminary—a calculation that would account for what was observed. He still believed that underneath it all, electrons really did exist, and moved in some clear ways. He'd limited himself to input/output descriptions simply because with the technology he had, there was no way to get more details. In the future, that would clearly improve.

Einstein was blunter with people he knew well. He always had been. During a stay at the lodgings of his successor at the German University in Prague, Philipp Frank, who had become a close friend, Einstein once politely corrected Mrs. Frank's inadequate attempt to fry liver in water, noting that fat or butter had a higher boiling point, and so would transmit heat more effectively. Ever after, the family called frying their meat in oil an example of "Einstein's theory." In one of their conversations, Philipp Frank made the same point that Heisenberg had: hadn't it been Einstein himself who had popularized the approach of just looking at external details? Einstein answered, sardonically, "A good joke should not be repeated too often."

To Michele Besso, Einstein was even more dismissive of Heisenberg's theories. Heisenberg's elaborate rules for transforming lists of what went into an atom into lists of what was observed coming out was, Einstein claimed, "a veritable witches' multiplication table . . . Exceedingly clever, and [yet] because of its great complexity safe against refutation as being incorrect."

Word of the august physicist's objections began to spread. Perhaps Einstein was right. Heisenberg, after all, was proposing a total shift in what everyone had believed. What if his Heligoland listing of inputs and outputs really was just a temporary gimmick—a computational shortcut—to be used until a better description came up?

IN THE PERIOD during which Heisenberg first confronted Einstein, the tables showed signs of turning in the elder scientist's favor. In January 1926, a graceful Austrian researcher, Erwin Schrödinger, had published a conventional, classical-style equation that, to many, no longer appeared to require relegating the movements inside an atom to the realms of unseeable mysteries. If his equation was correct, it seemed that it would return quantum mechanics to the strictly causal realm of physics that Newton and Einstein inhabited. Were that to be so, Schrödinger would be undermining Heisenberg's insistence that only a fundamentally new view—one that didn't even try to describe the inside of the atom in crisp, mechanical terms—could be accurate.

Heisenberg tried to fight back. But whenever he attempted to best Schrödinger in a debate, somehow he got flummoxed. Schrödinger was more than a decade older than Heisenberg and had a Viennese superiority and calm that drove Heisenberg to distraction. (He also had a private life that the Boy Scout–like Heisenberg could never grasp. Schrödinger had worked out his equation over Christmas 1925 at a luxurious Alpine resort—accompanied by one of the various mistresses his wife was happy for him to travel with—delicately placing a single pearl in each of his ears when he needed quiet.)

Erwin Schrödinger, about two decades after his 1926 breakthrough

Heisenberg had a dilemma: if you've made a great discovery and then been disbelieved, what do you do next? In desperation, he turned back to his most central belief. He was being criticized for saying that it was a waste of effort to try to trace the clear paths that electrons follow within atoms. Well, that's what he would face directly. He would go further than simply asserting one couldn't measure the behavior of those electrons; he would prove it.

Besides being scorned by Einstein and shown up by Schrödinger, Heisenberg had one other humiliation from his past that motivated him and had prepared him particularly well for this new challenge. Back in his student days, under Sommerfeld in Munich, he had been called to take his Ph.D. oral exams—the final step before obtaining his doctorate—at the unheralded age of twenty-one. Since Sommerfeld was the physics department's respected chairman and Heisenberg was his prize student, everyone assumed the oral exams would be a formality. But the faculty at Munich also included the elderly experimentalist Professor Willy Wien. Heisenberg had been registered for a course under Wien just before the exams, but he had skipped it almost entirely. He had never liked experimental work, he was excited about the forthcoming degree ceremonies, and anyway, he knew he was brighter than anyone else at the university. What could a harmless old experimentalist possibly do to hurt him?

Wien recognized he was no longer as respected as he had once been, and he'd also had a much harder life than Heisenberg—being raised on a rural estate that his parents had been forced to sell after a drought; repeatedly dropping out of school. He also believed that experimentation was the true basis of all advances in science. Sommerfeld, the theorist, had all the glory now, and Wien couldn't attack him—he was too powerful. Sommerfeld's student, however, would be different.

When Heisenberg entered the seminar room in the Theoretical Physics Institute at 5 p.m. for his oral exams, there Wien was, sitting beside a now slightly apprehensive Sommerfeld. Wien began the questioning mildly enough, asking Heisenberg how a certain new electronic laboratory device worked. Heisenberg didn't know. Sommerfeld tried to switch the topic, raising theoretical questions where Heisenberg's knowledge of mathematics would let him do well. Wien waited till they were done and then returned to his polite questions: Could Mr. Heisenberg perhaps now tell him how a radio circuit worked? Heisenberg tried to figure it out but then got lost, for these were details he'd never studied. Then Wien asked how an oscilloscope worked. Finally, Wien asked: Could Heisenberg even tell him how an ordinary microscope worked?

Heisenberg stumbled out of the seminar room two hours later, his face flushed, unwilling to speak to anyone. He told his father that his career in physics was over. Only the intervention of Sommerfeld—whose top grade for Heisenberg balanced Wien's equivalent of a failing F—allowed Heisenberg to get his degree.

That had been in 1923. Now, these few years later, after meeting Einstein in 1926, if there was one thing Heisenberg had gone over and over again in his mind, it was how to calculate how much a microscope could magnify what it was aimed at, and how exactly that process worked. This was the understanding he would use to show that no one could ever follow the detailed paths an electron took within an atom. It was also a good way to refute Schrödinger: "The more I think about the physical portion of Schrödinger's theory, the more repulsive I find it," Heisenberg confided to his friend Wolfgang Pauli later in 1926. And to his mentor Bohr: "I got the idea of investigating the possibility of determining the position of a particle with the aid of a gamma-ray microscope." Now he proceeded like no one had before.

If Einstein really wanted to see an electron, Heisenberg reasoned, he'd have to shine a light wave or some other energy down on the atom to light up the electron in there. But electrons are small. If the blast of light was strong, it would overpower the electron, jarring it out of position. Yet if the blast of light was weak, it couldn't be precisely aimed enough to see the tiny electron. It's much the way that, however carefully you use a gauge to measure the air pressure in a car's tire, you are inevitably letting a little bit of air out, so that the very act of taking the measurement makes your reading incorrect.

Heisenberg managed to prove that any super-microscope had to suffer from the same problems: it would be useless for observing the electron without influencing it. If you get a clean view of an electron's position, you'll knock it out of line with the light you're using to see it, and so won't be able to tell exactly what direction it had been traveling. (That's because individual packets of light carry a distinct momentum "punch" as they travel: it's very small, but enough to "push" a tiny electron.) But if you want to be so gentle that you don't knock it away from where it's traveling, then you won't have enough clarity to see exactly where it began. You can choose to measure either where the electron is or how fast and powerfully it's traveling, but you can't measure both with full accuracy at once. You're always going to be a little bit unsure—uncertain—about the complete mix.

This is the basis of the famous uncertainty principle, which Heisenberg published in February 1927. It was irrefutable. It ended centuries of belief that the universe followed an inherent perfect order. It revolutionized physics.

And Einstein would have nothing to do with it.

SEVENTEEN

Arguing with the Dane

THE DISAGREEMENT BETWEEN Einstein and most other quantum physicists first came to a head at the Brussels conference in October 1927. This was the same gathering where Lemaître buttonholed Einstein about the lambda. As if it weren't enough for him to be fending off one set of uncomfortable ideas, now he had two—and one fight would, in time, reinforce his determination in the other.

If the meeting had taken place just a year earlier, Einstein would have enjoyed the support of many of his assembled colleagues. Up until that time, many of the attendees had shared Einstein's initial responses to Heisenberg's ideas. Before Heisenberg's imagined work with a gamma-ray microscope yielded the uncertainty principle in early 1927, physicists were skeptical about his theories regarding the quantum universe. They, like Einstein, were impressed that his early computations had had such success in accounting for how electrons responded to blasts of light, but weren't convinced that reality could be so unclear, so vaguely glued together, that at the most detailed level we really had to accept uncertainty forever.

When it emerged in February 1927, several months before the conference, however, the uncertainty principle robbed Einstein of many potential allies. The principle, most physicists agreed, did seem to show that views into the atom were inherently closed off. Heisenberg, they conceded, appeared to have been right—which meant that Einstein (whom many of his colleagues would have known to be disdainful of the younger scientist's theories) had to be wrong.

Einstein had been invited to open the conference, since everyone wanted to see how he would deal with the new challenge from the quantum theorists, and how he would defend his traditional views about causality. He declined, however. He wasn't in a position to tell all of Europe's scientists what to think—not yet, and not in the magisterial way in which he'd been able to lay out the details of general relativity. His feelings were still a hunch, a suspicion, an almost visceral belief that "an inner voice" told him this could not be how the world worked.

So Einstein sat politely through the opening sessions and watched as Niels Bohr stood up to weigh in on the issue. Now middle-aged, Bohr had become the leader of the pro-Heisenberg faction. As he'd aged the odd appearance he'd had as a young man had become more attractive. His habit of speaking slowly and softly, with long pauses for thought, gave his words majesty.

Bohr began the conference by recapping the changes that Europe's scientists—there were almost none of any significance at the time in the United States—had been experiencing. Since the decline of medieval scholasticism, Bohr recounted, there had been at least some efforts in the West to bring reason to bear on the material world. This was not a fettered reason, predetermined to come up with conclusions that matched what the church wanted to hear. Rather, this was a reason, an intellectual inquiry, that was convinced it could uncover every fact of nature, however laborious the process might be and however many centuries it might take. This program of inquiry took for granted the belief that what was out there, in the real world, truly existed and could—in whatever detail we wished—ultimately be understood.

That certainty about the universe was what these new findings seemed to be undermining—in Bohr's view quite definitively. Causality of an absolute, classical sort didn't exist. We might think there were exact sequences of events that must follow one another—kick a soccer ball hard, for instance, and it will spring forward—but that's just because we're seeing the averaged results of a vast number of submicroscopic encounters, each one operating by chance. The electrons on a player's boot get very close to the electrons on the leather surface of a soccer ball as he swings his leg forward. That we can see; that we can know. But which of those electrons will repel one another, sending the ball flying away, can never—not even in principle—be entirely known.

The uncertainty principle proved that these subatomic goings-on were unknowable, Bohr insisted. The micro-world really was different from the ordinary large-scale world we're used to. On the smallest scale, chaos and indeterminacy ruled how the electrons and other particles that make up our bodies and planet operated. Clarity at the micro level did not exist.

Einstein had come to know the shambolic, brilliant Bohr very well over the years. At their first meeting, in Berlin in 1920, Bohr had brought along Danish cheese and butter, which was much appreciated in a city still suffering from the recent British blockades. At another meeting, in Copenhagen, they'd been so caught up in conversation—admittedly, much of which must have been Einstein waiting as Bohr paused to put together his intense whispers—that they'd far overshot the streetcar stop for Bohr's home, turned around, and then overshot the stop coming back as well. They were the two wise men in their field, and they also liked each other. "Not often in life has a human being caused me such joy by his mere presence," Einstein once wrote to Bohr. He wasn't going to insult his old friend by mocking this most fundamental of new beliefs in public.

Only outside the main sessions, after Bohr had publicly said his piece, did Einstein begin to argue back.

• • •

Einstein and Bohr in a reflective mood, probably at their friend Paul Ehrenfest's house, late 1920s

BOHR PRESENTED A SHAMBLING FRONT, and took even longer than Einstein to get a pipe lit and keep the tobacco burning. (He carried an extra-large box of matches with him to help.) But he was committed to physics, and in a certain sense to the cause of Professor Niels Bohr. His parents were prosperous and distinguished, and with the confidence from growing up with their connections, he'd arranged—despite the lumbering appearance, he was the most skilled of bureaucratic operators—for the Carlsberg Foundation to support a great research institute under his direction in Copenhagen. Through scholarships, grants, and publications, that institute was doing everything it could to support the view that Bohr took of Heisenberg's and Born's results. It would be embarrassing if he were to be proved wrong. Instead of the leader of a new breakthrough in thought, he would look like a middle-aged professor who had jumped on the latest bandwagon simply to seem up-to-date.

It still seemed possible, however, that Einstein would be able to prove Bohr wrong. All Einstein had to do was show how to construct one machine that could operate in contradiction to the uncertainty principle. If he did that, Bohr's support of the uncertainty principle would be shown to be empty. The possibility that Einstein could accomplish this was very real. Einstein was, after all, the man whose thought experiments about a falling elevator had led to startling yet fully accurate predictions about starlight swerving near the sun; who had, in 1916, as just one of the more minor of his other thought experiments, envisaged a machine that could amplify light on call—the machine that would eventually become our laser. Who was to say that he could not solve this puzzle, too?

Yet Einstein, like Bohr, had a lot on the line. At forty-eight years old, Einstein knew that he was nearing the point where physicists often go from creating fresh ideas to disparaging whatever is new. He had certainly been on the receiving end of that latter disposition when he was younger. His whole self-definition depended on not being like that. He was a revolutionary. He thought independent thoughts; he went wherever the truth led; he didn't want to be constrained by the heavy bourgeois style of the Berlin apartment he lived in with Elsa, or by the worst of her social-climbing friends. He had made his own light and airy refuge up in the attic; he wore loose sweaters and often went barefoot around the house, regardless of whether visitors thought such casual behavior beneath him; he was limited only by what he understood as the minimal true structure of the universe.

What he needed was one successful construction. It didn't even have to be built; it would be enough if he could describe it in words and show Bohr and Heisenberg that it worked. If he could do that, he'd be back where he knew he belonged—at the forefront, consolidating the truth, not anxiously trying to hold on to the past simply because that happened to be what he was familiar with. And he knew in his bones that the universe had to have causality in its deepest structure; he was convinced of it. How, then, to show that was true?

It helped that he could make almost any mechanical device work. He'd had years of practical experience analyzing the most complex of devices in his Patent Office days. That would be his approach here.

Heisenberg later recalled how Einstein went about launching his attack. They were all staying at the same hotel, he said, and Einstein had the habit of telling the others at breakfast about experiments he had thought up that would, he felt, undermine quantum mechanics. As Bohr, Einstein, and Heisenberg walked to the conference hall together, they would make a start on analyzing the assumptions behind Einstein's latest proposal. Heisenberg picks up:

"In the course of the day, Bohr, [Wolfgang] Pauli and I would frequently discuss Einstein's proposal, so that already by dinner-time we could prove that his thought-experiments were consistent with the uncertainty relations, and so could not be used to refute them. Einstein admitted this, but next morning brought along to breakfast a new thought-experiment." Each time, the new thought experiment was more complicated than the previous one, but each time —by dinner—the other men had managed to disprove it. "And so it went on for several days."

Einstein's close friend Paul Ehrenfest, from the Netherlands, was also at the 1927 conference, and shortly afterward he told his Leyden students about it. He loved listening to the dialogue between Bohr and Einstein. Einstein "was like a chess player," he felt, coming up with ever new examples. "He was a perpetual motion machine, intent on breaking through uncertainty." But there also was Bohr, who "out of a cloud of philosophical smoke" would lean forward, musing and musing until he came up with the tools that could undermine Einstein's new examples. Sometimes, when Einstein had devised an especially puzzling "demonstration" of why quantum mechanics had to be wrong, Bohr would keep Ehrenfest up nearly all night as he thought out loud until he found the flaw.

• • •

THE CONFERENCE ENDED in a draw. Einstein had failed to find a counterexample that would refute Bohr, but Bohr remained apprehensive that this new theory on which he'd staked so much might still be undercut.

On the way back to Berlin, Einstein consoled himself with the thought that the argument wasn't simply one of youth against age, with all young physicists on Heisenberg's side and only old ones on his own. It helped that he shared the first part of the journey, to Paris, with Louis de Broglie, a dignified French physicist a decade younger than him, who'd done fundamental work laying out the principles behind quantum mechanics, yet who had the same doubts Einstein did. De Broglie too was convinced that Heisenberg's explanation was just a provisional step and that somehow a core of certainty would eventually be found underpinning everything we saw. (De Broglie had personal reasons to feel benevolent, for Einstein had ensured that his Ph.D. dissertation, in which he had set out those ideas, had been accepted.)

The calculated results from quantum mechanics that Heisenberg and others had come up with were quite accurate, both Einstein and de Broglie agreed, but as Einstein repeated, "I believe that the limitation to statistical laws will be a temporary one." On the Gare du Nord platform in Paris, engaged in one of those long talks travelers have when neither wants their shared journey to end, Einstein repeated his points. De Broglie agreed, and as he left, Einstein called after him, "Carry on! You're on the right track!"

In the two years following the 1927 conference, however, Einstein began seeing that his side in the quantum debate was losing popularity. More and more experimental demonstrations seemed to show that quantum mechanics worked. De Broglie himself only held out till 1928 before joining the consensus that Bohr, Heisenberg, and the others on their side must be right. It was becoming a trend. The Austrian Erwin Schrödinger, soon to receive the Nobel Prize, was one of the few scientists to remain on Einstein's side.

By 1929, however, Einstein had good reason to be more confident, despite his diminishing support. He was genuinely a modest man, who knew his intellectual gifts weren't as extraordinary as the general public believed. Grossmann in Zurich, Born in Göttingen, and many others were stronger mathematicians. If he, Einstein, did have good physical insights, that was because his family had raised him in such a distinctive way: open-minded enough to be critical of received opinion, yet grounded in the solid reality of lightbulbs, electric generators, and all the other whirring apparatuses his father's and uncle's income had depended on. Lurking behind his insights may also have been his ancestors' only semi-forgotten religious beliefs, and especially the assumption that there had to be a waiting order and certainty, which at selected moments we were lucky enough to see. And from that mix, of which he had merely been the lucky beneficiary, he also knew that he *had* been able to probe beyond surface appearances to underlying principles that only much later had experimentalists found to be true.

Einstein's $E=mc^2$ equation was now almost universally accepted. But there was something even better. During the 1927 conference, and despite Lemaître's claims, it had still seemed likely to Einstein that the lambda addition to his other great equation was necessary: that astronomers had been right and his magnificently pure $G=T$ had to be discarded; that his belief in the power of sheer intuition was wrong. But just this year, in 1929, Hubble and Humason had published their new work showing that Einstein's original, beautiful equation had been right after all.

To Einstein, Hubble and Humason's findings changed everything. What they'd discovered with their great 100-inch telescope—that the lambda term wasn't necessary—showed that his original intuition there, too, had been right—that what he'd seen in 1915 about "things" altering geometry, and altered geometry guiding "things," had been absolutely, 100 percent true. Experimental results—all the assumptions of the world's astronomers—had seemed to show otherwise, but if Einstein had held out, he would have been proved right.

Clearly, he believed, he could hold out—and be proved right—again. He had already been disposed to believe that the universe had to be fundamentally knowable. His experience with lambda—showing that his initial intuition had been justified—provided an extra boost.

Admittedly there was a great danger here. The English essayist Macaulay once said of himself—accurately, if not modestly—that he had an excellent writing style, but it was close to a very bad style indeed. This meant, he warned, that few of his readers should try to copy it, for if they got it even a little bit wrong, they would fail entirely. Einstein increasingly was taking a similar risk. Advancing from a belief that his intuition was right is what had made him the greatest scientist of the modern era. Yet holding *only* to that approach meant that his self-confidence could easily cross the line to sheer dogmatism. What's more, he was less constrained than ever in these issues. During his university years in Zurich he'd had to be responsive to the best wisdom of the past, and during the years with Grossmann he'd had to defer to a friend's superior mathematical talents, but now he found himself unshackled from these constraints—and more than a little untethered.

Unless, of course, Einstein really was right. No one yet knew for sure.

THE WORLD'S TOP physicists only assembled in Brussels every few years. Since the 1927 conference had ended in a draw, when the next one arrived, in October 1930, everyone's attention was on Einstein and Bohr. They were the two intellectual giants of their generation. Would they clash again, as they had at the last meeting?

Einstein knew this was the last opportunity he would have to keep the community of physicists on his side, especially the young generation, with whom he'd identified for so long. Yet in 1930, as at the previous conference, he remained quiet in the main meetings; once again he would only bring his objections to Bohr in the relative privacy outside those plenary sessions. In the meantime, the Dane worried.

Bohr knew something big was coming, but how could he prepare? He simply had to believe that the newly developed science of quantum mechanics would be strong enough to stand against anything. Heisenberg steeled himself, too. Like chess grandmasters before a match, he and Bohr and others had tried to plan every defense.

Einstein, too, must have spent a long time preparing, puffing on his pipe in his Berlin study or at his country house, for what he came up with was tremendous.

At the heart of quantum mechanics was Heisenberg's uncertainty principle, which seemed to put a limit on the detail we could hope to see on the micro level. Without that detail, we could never be sure, entirely, just what was going to happen next. Heisenberg had first presented his principle as saying that one couldn't get complete accuracy in measuring a particle's momentum and position at the same time. It was, as the future Nobel laureate Wolfgang Pauli put it, as if we could see an object's momentum by looking out our left eye, and its location by looking out of our right eye, but would be stuck with a blur if we tried to keep both eyes open at once.

Previous attempts to get around Heisenberg's principle had failed for the same reason that attempts to use a tire pressure gauge fail to give fully accurate readings: the very act of using the gauge lets air hiss out, and so changes the pressure inside the tire that you're trying to measure. Einstein's new idea was to step back and view the "tire" from farther away: not using any sort of gauge or other device that would disturb it.

Einstein's approach was akin to simply weighing the tire, instead of measuring any air going out of it. He came up with a way to do this because recent work had also shown that Heisenberg's principle meant that one could measure a particle's energy or the exact time at which it had that energy, but not the two at once. This new finding about the uncertainty principle allowed Einstein to mount the most vigorous attack against it yet.

For his new thought experiment in Brussels, Einstein came up with a device that would have done his old Patent Office supervisor Herr Haller proud. After they'd strolled away from the main meeting sessions, he told Bohr to imagine a box that had a fine cloud of radiation—think of it as a cloud of light particles, or photons—floating inside it. There's a tiny shutter in one wall, controlled by a very precise clock. The whole apparatus is supported on a scale, so it can be weighed. When the clock strikes a particular time, the shutter opens, one photon is let out, and then the shutter closes. The box is weighed before and after, and that way it's obvious how much mass has been lost.

Doing this, we know how much energy that lost photon carries: the scale tells us (because mass and energy are equivalent). We also know what time it is when the photon flies out: the clock tells us. This was something that should never happen if Heisenberg's uncertainty principle were true. Since the clock has no connection with the scale—unlike a tire pressure gauge, where measurement interferes with accuracy—Heisenberg's argument is ruined. Certainty is possible. The classical world of cause and effect is saved.

Bohr knew that he thought more slowly—albeit more deeply—than most others. But he was used to getting at least some feeling of what the solution to a problem might be. For Einstein's light-filled box, however, he could imagine no solution at all. The photon flies out through the shutter. The clock records the time. The scale moves. The clock and the scale are nowhere near each other.

How could that be reconciled with Heisenberg's uncertainty?

Einstein's thought experiment overwhelmed Bohr. As one contemporary recalled, "[Bohr] was extremely unhappy, all through the evening, walking from one person to another, trying to persuade them all that this could not be true . . . But he could think of no refutation. I will never forget the sight of the two opponents leaving the university club: Einstein, a majestic figure, walking calmly with a faint ironic smile, and Bohr trotting along by his side, extremely upset."

It was Einstein's last moment of glory. Bohr stayed up almost all night—no doubt dragooning graduate students or anyone else unlucky enough to be nearby for help—as he tried to mumble his way to a solution. Heisenberg had earlier described the way that once Bohr was focused on a problem, "he would not give up, even after hours of struggling." So it was here.

In the morning, Bohr had it. When the shutter opens and the photon flies out, the mass of the box goes down. But the weight of the box is being measured. That means it has to be on a scale. When the photon flies out, the scale rises up—very little, but at least a bit. That means it's ever so slightly higher in the earth's gravitational

Einstein and Bohr at the 1930 Brussels conference, photographed by Paul Ehrenfest, probably the day Einstein proposed his box+clock experiment but before Bohr had analyzed it.

field. By Einstein's own theory of relativity, time is seen to operate at different rates in a stronger versus a weaker gravitational field.

Bohr sketched out the calculations, and once everyone staying at the hotel saw where this was going—Bohr, Heisenberg, probably Ehrenfest, and perhaps others—Einstein, to his credit, helped them fill in the details. Working together, Einstein and Bohr concluded that the uncertainty in the weighing, because of that tiny gravitational shift, was just enough to match exactly what's predicted by Heisenberg's uncertainty principle.

Einstein had neglected his own theory of relativity—and Bohr had used it to refute this final attempt to defend causality. It was a crushing blow, made all the more painful by the fact that it was delivered by Einstein's own instrument—and its implications could not be clearer. Back in 1916, Einstein had assumed that using probabilities to describe how photons operate within a device such as his proto-laser was just a temporary measure and would be put aside once science went further and our knowledge increased. That dream was now over.

Heisenberg was exultant at the outcome. When he saw Einstein's last bastion crumble, he wrote, "We . . . knew that we could now be sure of our ground . . . The new interpretation of quantum mechanics could not be refuted so simply."

Bohr was the humbler man, but the gist of his polite, guttural mumblings was clear: He had won. Einstein had lost.

INTERLUDE 4

Music and Inevitability

EINSTEIN NEVER AGAIN attended such a meeting; never again attempted to refute Bohr or Heisenberg in public debate. Nor, however, did he change his beliefs. He was still convinced the world's experimentalists were wrong, their findings incomplete.

For his consolation, he turned to music, as he always had. Einstein loved much of the classical repertoire, even if he took issue with most of its composers. "I always feel," he wrote, "that Handel is good—even perfect—but that he has a certain shallowness." Schubert failed the ultimate test, too. "Schubert is one of my favorites because of his superlative ability to express emotion and his enormous powers of melodic invention," Einstein admitted. "But in his larger works I am disturbed by a certain lack of architectural shape."

The imperfections went on and on. "Schumann is attractive to me in his smaller works," Einstein wrote, "because of their originality and richness of feeling, but his lack of formal greatness prevents my full enjoyment . . . I feel that Debussy is delicately colorful but [also] shows a poverty of structure." In conclusion, he wrote, "I cannot work up great enthusiasm for something of that sort."

How could these otherwise great composers have missed the large-scale unity that he knew was out there to be found? Only Bach and Mozart had accomplished that. Those two had something that surpassed the others. "It is impossible for me to say [which one] means more to me," Einstein wrote, but what he knew for sure was that no one else measured up. Beethoven might have been expected to be at that top level, for instance, but though Einstein found him powerful, Beethoven was also "too dramatic, and too personal." His work had something arbitrary about it, for human emotions depend on our bodies and our personal histories. Mozart, however, went beyond the realm of personal emotions, with a music "so pure that it seemed to have been ever-present in the universe, waiting to be discovered by the master." Mozart's work felt more "necessary," letting us see a Platonic realm of truth that exists far beyond the chance events of anyone's personal history.

Einstein sought in the music of Bach and Mozart precisely what had eluded him elsewhere. In his emotional life, in his marriages, even more so in his affairs, Einstein had failed to find anything lasting, anything certain. His failure hurt all the more because his dream of certainty, and of contact with the truth, still haunted him.

In letter after letter now, he went over the many ways in which his previous work had ostensibly proved that his beautiful dream was valid. $E=mc^2$ from 1905 showed there was certainty in the universe, since it described in as much detail as one could wish exactly how mass and energy could change into each other. The great G=T of his 1915 equation had been just as clear. Mass made space curve. Curved space guided mass along. How could there be any random chance involved, given that this equation, too, was so clear? The sheer simplicity of G=T was impossible to ignore. "Hardly anyone who has truly understood it will be able to escape the charm of this theory," Einstein had written, exhausted but content, that Berlin winter when he'd first completed his work on the equation. He himself was still trapped within its orbit.

It's true that Einstein himself had once questioned the simplicity at the heart of G=T—during the years between 1917 and 1929, when his lambda mistake still stood—but ultimately this questioning had proved unnecessary. Moreover, although Einstein may have been humiliated in Brussels in 1930, he also took heart in the fact that his contemporaries had validated his other work time and time again, in ways that lent support to his belief in the inherent certainty of the universe. Humason had measured distant galaxies through the giant telescope in the mountains of California and found that billions of stars were hurtling away from us. There was no ambiguity about it, and it was exactly what the original, simple G=T predicted. Such reinforcement helps explain why almost a decade after the 1930 conference, Einstein still felt comfortable telling a close assistant, "When I am judging a theory, I ask myself whether, if I were God, I would have arranged the world in such a way."

EINSTEIN'S FAITH IN his own ability to judge the architecture of the universe was powerful, but also potentially dangerous. The more esteem a great man gets, the easier it is for him to deny reality—just as Einstein was now doing, and in a manner of which his younger self would have disapproved.

Einstein had once drawn for his old friend Maurice Solovine—the enthusiastic Romanian who'd first responded to his 1902 advertisement for private lessons in math and physics in Bern—a picture of how he felt creativity worked. We start with the reality around us, Einstein wrote: the empirical world, where we experience our ordinary sensations. In a burst of imagination, thinkers can ascend from that foundation to loftier general principles. Then, in order to be sure that those principles are true, we must work out detailed propositions that follow from those principles and test them against the empirical world.

That was the procedure Einstein had followed with $E=mc^2$, whose predictions—after he'd conceived of them on paper—he'd proposed testing with the radium salts the Curies were using in Paris. It was the procedure he had followed with general relativity as well: a great jump in imagination—using the thought experiments about the falling room—to create a clear, abstract theory, and then from that drawing detailed testable conclusions, such as the ones about space curving that Eddington had checked during the 1919 eclipse.

Although Einstein often wrote that this was still proper, he also increasingly expressed a contrary belief. As he wrote in 1938 to an old colleague, "I began with a skeptical empiricism . . . But the problem of gravitation converted me into . . . someone who searches for the inky reliable source of Truth in mathematical simplicity." As his work went on, Einstein increasingly ignored his original, more empirical approach. "[Quantum theory] says a lot," he wrote, "but does not really bring us any closer to the secret of the 'Old One.' I, at any rate, am convinced that *He* is not playing at dice." God, he felt sure, followed a rational plan when He designed the universe. Experimental results were not going to rebut that.

Nothing at the Brussels conferences, apparently, had changed his mind.

His whole belief system would have been crushed if something had. But when he said "God does not play dice with the universe," Niels Bohr would reply, effectively, "Einstein, stop telling God what to do!" The two men held utterly different views—not just about how the universe worked, but also about their own abilities to discern its divine functions.

Only one of them could be right.

Part VI

FINAL ACTS

Einstein in Princeton, early 1950s

EIGHTEEN

Dispersions

BY 1950, TWENTY years after the final Brussels conference, Bohr's institute in Copenhagen was at the center of the world's physics research. Despite his victory over Einstein in 1930, the towering Dane had managed to avoid the lure of dogmatism, and his broad-mindedness had attracted some of the brightest new minds. Young people from Harvard, Caltech, and Cambridge eagerly went to Copenhagen for a year or two during their graduate studies or after, to join in the exciting atmosphere and share ideas with the respected, approachable Professor Bohr. Conversations with him demanded as much concentration as ever, for Bohr's accent still rarely strayed far from Danish no matter what language he tried to speak. But that didn't matter. The bright young people at the institute came from so many countries that they happily described its official language as "Broken English."

Bohr was a hero in his own country. After the outbreak of World War II, he'd kept the institute operating during the first years of the German occupation, staying until 1943, before being spirited off—by secret RAF transport from Sweden—when his Jewish ancestry and political significance made staying any longer too dangerous. Excessively tall, and excessively polite, Bohr had nearly died during the RAF flight, for he'd been secreted away in the bomb bay and was supposed to speak into a microphone to tell the pilots if anything was wrong. When his oxygen failed—the mask not fitting around his head—his mutterings and polite gasps seemed no less incomprehensible than his previous communications, and he fell unconscious, only recovering when the pilots, realizing it was odd that there was such silence, swooped lower into denser atmosphere, where there was enough oxygen to keep Bohr alive.

Brought in to aid the Manhattan Project in building an atomic bomb, Bohr had tried, albeit unsuccessfully, to alert both Churchill and Roosevelt to the dangers this weapon posed. He suggested there should be a demonstration of it first, or arrangements set up for international control, but to no avail. When the United States dropped the bombs on Hiroshima and Nagasaki in the final days of the war, it was the first time the world had been treated to a public showing of these terrible machines—weapons that had been born, ultimately, in the theories of Einstein as much as in the practical efforts of Bohr and many others.

Bohr always felt, as he once put it, that we should be both "spectators and actors in the great drama of life." With the support of his thoughtful wife, the openness of his personality, and the general safety of Denmark, he'd managed to be both an onlooker and a participant—in politics as well as in science—all while keeping in line with Europe's noblest ideals. He emerged from the conflict not only unscathed but stronger in his public standing, at least, than ever before.

The German physicist Werner Heisenberg, by contrast, had disgraced himself during the war. More worldly physicists had sometimes teased him for his years of rambling over the countryside with exuberant youth groups. But those hikes hadn't been as innocuous as they'd seemed. Ever more of their participants felt this was a way to get close to the Fatherland's soil and to help preserve it from dangerous outsiders, such as Jews and foreigners. Although Heisenberg did try to stand up for a few of his colleagues who were being ousted from their academic positions for being Jewish, later he clearly relished being brought into commanding positions within the technocratic parts of the new Nazi state. There were new research groups to run, large budgets to control, and visions of a miracle weapon that could ensure Germany's triumph over its enemies forever.

At one point during the war, with black-jacketed SS officers not far away, Heisenberg had even stormed into Bohr's institute in Copenhagen, explaining with great assurance now—in these early days, when Germany was on the rise—on which side the future lay. Bohr was still there, and he was appalled. Already he'd begun preparing the institute against German depredations, including hiding the gold Nobel Prizes of two Jewish members. (By German law, what was owned by Jews could be stolen, and if the owners tried to keep their possessions, such as by shipping the medals abroad, they or anyone who helped them could be arrested and quite legally tortured.) Bohr's ingenious friend de Hevesy from their Manchester days, now in Copenhagen, had worked out the ideal hiding place. The lustrous gold medals were dissolved in a mixture of nitric and hydrochloric acids, creating an innocuous brown sludge that was stored on a back shelf until the war was over.

That was the German state the excited Heisenberg now so happily represented. His brilliance in creating the uncertainty principle had given him the respect within the Nazi establishment to do almost whatever he wanted. Bohr didn't know that Heisenberg would soon be working female slaves to death at the Sachsenhausen concentration camp by forcing them to produce toxic uranium powder for his experiments. But Bohr was a civilized man. He now recognized, with disgust, that Heisenberg, despite his music, his education, his mathematical brilliance, was not.

Heisenberg's old teacher, Max Born, being Jewish, could not ride out the war like his pupil; he had to escape Germany. Even at the time of the 1930 conference, the youth groups that Heisenberg relished had been getting stronger, and in the quiet university town of Göttingen, about a third of the adults voted for the Nazi Party in elections that year. One especially energetic student group began working through baptismal records and town registries to see which professors were actually Jewish. Detailed lists were drawn up, and *The Jewish Influence in German Universities,* vol. 1, *University of Göttingen* appeared. Just a few years later, lists like that would be used for extermination.

Life became impossible for Born, especially as almost all his faculty colleagues turned away when he tried to get their support. Eventually he ended up in Scotland, where he became a benevolent teacher of generations of students. (His daughter, marrying a British man and taking his last name, Newton-John, later moved to Australia, where one of their own children, Olivia, had noted success as a singer and actress.) It was a good thing he got away when he did, for during the rise of the Nazi state, it became clear that intellectuals and other prominent Jews were under particular threat.

In 1933, when Born was still in Germany, Hitler gained effective control of the Reichstag, and the great number of students who were Nazi supporters could beat up Jews with impunity. Born's daughters were threatened on the street. And then, on May 10—in a scene unimagined since the Middle Ages—throughout the country, including the old university towns, great pyres were made of books.

The largest book-burning crowds assembled in Berlin at the Opernplatz, just near the Opera House. Students had eagerly been collecting cartloads of volumes seized from libraries or private homes. Propaganda Minister Goebbels arrived at midnight to begin a nationally broadcast speech: "German men and women! . . . You do well in this midnight hour to commit to the flames the evil spirit of the past!" Goebbels's photographers were standing by, ready to capture the images that would be shown across the country: the joy before flames, the exultation in the crowds. Student crowds in Göttingen had engaged in their own burnings the same night.

Einstein's books had been hurled into the flames with especial glee, for he was the most famous of all Jewish intellectuals and represented a spirit of liberalism and rational inquiry that was the opposite of what the new state insisted was right. "The age of Jewish intellectualism has come to an end!" Goebbels announced to the nation from Berlin's Opernplatz. It was easy to tell what was coming.

LATE IN 1932, the year before the Opernplatz rally in which his books would be burned, Einstein had gone to his country home outside Berlin with Elsa—that place of the affairs that had tormented her; of the friendly walks and mushroom hunting and family dinners she had loved. Now they were there to collect his papers, as well as her most important belongings. Caltech, in Pasadena, California, had offered him a position, while Princeton's new Institute for Advanced Study in New Jersey looked set to propose a better one.

Elsa was good at reading people, but her intuition failed her about what was happening to her country. She and Einstein had gone to America before, for visits or even longer stays when he was a months-long lecturer. Surely this would be just the same?

Einstein shook his head. She understood very little. "Look around you," he reportedly said. "It's the last time you'll see this."

After Einstein and Elsa had left their home, and after the book burnings the following year, mobs broke in, ransacking what possessions the hated Professor had left. Elsa only found out about that later. She was in Belgium at the time, having been under armed protection with her husband before they sailed to America.

NINETEEN

Isolation in Princeton

EINSTEIN SPENT THE rest of his life, from 1933 to 1955, in Princeton, a university town then far from the sophisticated, egalitarian haven it has become today. There were few Catholics, fewer Jews, and no blacks allowed to teach or attend when Einstein was first there. The faculty thought highly of themselves, even though for most of them the prestige that Princeton afforded was not close to that which they would have enjoyed at the genuinely important institutions of the day—such as those in Zurich, Berlin, or Oxford—which unlike Princeton were home to world-class scientists who produced essential work. Faculty parties were especially ridiculous, and certain professors put on airs that even Elsa's socialite friends would have found excessive: making blue-collar New Jersey men dress up as liveried footmen and bow as they served champagne on fine trays. Writing to a friend in Belgium, Einstein described the whole setting as "a quaint and ceremonious village of puny demigods, strutting on stiff legs."

Ordinary residents in the New Jersey town were more pleasant. When the great black American singer Marian Anderson was refused entry at a local inn, Einstein invited her to stay at his house and found that instead of being ostracized, a number of his neighbors quietly supported him. They liked this pleasant European in their midst. Indeed, on his very first day in Princeton, Einstein entered an ice cream shop and, knowing his barely understandable English would not get him far, had jabbed one thumb toward a student with an intriguing carrying device for the ice cream, then pointed to himself. The waitress who served him his first vanilla ice cream cone later told reporters it was one of the highlights of her life. The fact that Einstein then strolled outside and bought a newspaper describing the way American journalists were hunting for news of his whereabouts (he'd been brought by tugboat directly from his ocean steamer to a pier in Manhattan, then whisked to Princeton to avoid publicity at the main Manhattan docks) simply added to his charm.

As time went on, he tutored neighborhood children in mathematics; at Christmas he went outside to play his violin with carolers; he bought a boat for holidays—a small seventeen-footer that he solemnly named the *Tinnef* (Yiddish for "piece of junk") and in it again he'd drift, content, for hours. He and Elsa were still not quite in love but made a decent shared life in this land of the flying snakes. When she experienced eye and then kidney problems, she wrote to a friend, "He has been so upset by my illness . . . I never thought he loved me so much. And that comforts me."

The material comforts were pleasant, too, as even in Berlin the Einsteins had not had an electric refrigerator. Here seemingly everyone had one. It was also—a great joy—easy to heat water for the bubble bath he liked in the morning. And with rural New Jersey all around them, the price of the two sunny-side-up eggs he preferred for breakfast was exceptionally reasonable. "I have settled down splendidly here," Einstein wrote his old friend Max Born. "I hibernate like a bear in its cave, and really feel more at home than ever before in all my varied existence."

But Einstein was hibernating in other ways as well. Where once he had walked the delicate line between stubbornness and suppleness, now he was becoming increasingly closed-minded. In his view, of course, there was no alternative. "I still do not believe that the Lord throws dice," he noted, even after several years in America. "Because if he had wanted to do that, he would have done it throughout, and not kept to [any] pattern. Gone the whole hog. In [such a] case we wouldn't have to look for laws at all."

His friends back in Europe begged him to reconsider his position. Every new discovery was backing the subatomic interpretations of Heisenberg and Born; there was absolutely no evidence on his side. The research indicated that scientists could study the world in ever finer detail, but there wouldn't be any certainty, guarantees, or determinism at its very core. Instead, there would be an intrinsic blur, an uncertainty—actions that seemed impossible from our large-scale perspective.

Einstein insisted those findings were just temporary and one day would inevitably be overturned. Yet in closing himself off from all that supporting data—which he found so repugnant to his view; which he felt the whole lambda interlude justified him in ignoring—Einstein also was closing himself off from the intellectual connection he still sought. For although much of the Princeton faculty was merely pompous, there were several researchers with whom he might have done serious work, much as Bohr was doing in Copenhagen.

Just blocks away from Einstein's Institute for Advanced Study, for example, in Princeton's main physics department, work was taking place on what would later be called quantum tunneling. Place an electron in front of a wall, and according to traditional physics, the electron might wobble around a bit, but otherwise would pretty much have to stay in its place. With the insights codified in Heisenberg's uncertainty principle, however, measuring the electron's velocity requires it to have an indeterminate location, since any measure taken of an electron's speed prevents an accurate reading of its location. What this means is that although there's still a chance the electron might stay in front of the wall, there's also a chance that when you next look, it will appear on the far side of the wall without ever passing through it on the way.

Were such quantum effects to be noticeable on our usual, large scale of existence, everyone would be able to walk through walls, be they brick or metal or stone. Thin steel walls would be easy to traverse, the walls of King's Cross train station in London would be a bit harder, and teleporting across the Matterhorn by running into its side would be left only to the most adventurous. None of that would be a matter of just pushing through the barrier in question. Rather, if the rules of quantum tunneling applied at this scale, first you'd be on one side of the thing, and then, instantly, you would appear on the other side.

By Einstein's intuition, this was impossible. Yet according to the data accumulated by researchers following Heisenberg, Bohr, and Born, it was what *did* happen in our real world. The men in the physics department at Princeton who were sharing this work worshipped Einstein and would have relished the chance to collaborate with him. Their studies ultimately helped lead to the creation of the transistors that today operate inside all our phones and electronic devices. But Einstein couldn't bring himself to grapple with these strange consequences of the new quantum mechanics. Quantum tunneling—and the transistor revolution—advanced without him.

EINSTEIN'S PERSONAL HISTORY had made him predisposed to discover relativity, but not to accept uncertainty. And now, like so many famous individuals—feted, financially free, his oldest friends far away—there was no force pushing on him to make him reconsider.

Instead, now in his fifties, Einstein began to concentrate more and more on what he termed his unified field theory. The great Victorian scientists had managed to bring much of what was known about energy in the universe together, fusing that knowledge in the concept of the conservation of energy, which held that all energy—

whether produced in a gas explosion or a slamming car door—was connected and could not be created or destroyed, but rather only transformed. In 1905, with $E=mc^2$, Einstein had taken that idea further, showing that not just all forms of energy but also all forms of mass were interlinked. In 1915, with $G=T$, he had shown that the very geometry of space was interlinked with the mass and energy held within all "things" as well.

Einstein had advanced the field of physics more than anyone in living memory. But what if he could go further and show that electricity itself was just another aspect of gravity and geometry? That truly would be an achievement for the ages and help show his critics that clear, causal links could be found between an even greater range of phenomena.

Such, at least, was his aim behind unified field theory, though here again his stubbornness worked against him.

When Einstein had been an undergraduate in Zurich, his professor Heinrich Weber had said, "You are a smart boy, Einstein, a very smart boy. But you have one great fault: you do not let yourself be told anything." Far from a great fault, at the time Einstein's bullheadedness was a strength, for Weber was locked in the physics of the mid-1800s, and Einstein needed to revolt against teachers like him in order to achieve greatness. Yet now, as an aging man, what began as a foible—if that—had grown into something more serious.

By staying away from the latest findings in quantum mechanics, Einstein was also isolating himself from the era's breakthroughs in recognizing new particles within the atom. For any unified field theory to work, it would have to incorporate those findings; it could not possibly succeed without them. Once, Einstein would have acknowledged this; indeed, he had regularly ended his early papers with a call to judge them according to new experimental evidence. Now, not only did he not call for experiments to test his theories, but, with the unified field theory being so far from what any researchers were engaged in, such a thing was not even possible. He wasn't responding to new research results; he wasn't proposing new, detailed experiments. His dream of a unified theory was proving impossible to carry out.

Einstein's insistence on his own path wasn't heroic self-belief anymore; it really was unreasonable stubbornness. Yet with his dogged determination, he kept at it, month after month, for almost twenty years.

His efforts were made more meaningless by the fact that working so much on his own now—or just with bright but utterly subservient graduate assistants—Einstein was also isolating himself from fresh analytic tools. A young visitor to his upstairs study saw that his work surfaces were full of papers still using the notation that had been so useful when Grossmann had taught him about it in the 1910s. By the 1940s and 1950s, physicists were using very different formalisms for their new work in nuclear physics. Yet those old tools had done such wonders for Einstein before that he couldn't let them go.

This was a tragedy, for Einstein's intellect remained exceptionally powerful. Several years into his time at Princeton, when he briefly set aside his work on unified field theory and turned back to pure relativity, he elaborated on a magnificent construct called gravitational lenses, suggesting that entire galaxies could warp everything around them so strongly that light from galaxies even farther behind them—light that should have been forever blocked from our sight—could actually be seen, as the warping "tugged" that light all the way around. The idea was so staggering that it was almost entirely ignored.

Amidst these other projects, Einstein summoned the energy to mount a final challenge to his bête noire, quantum theory. In 1935, working with two younger colleagues, he tried one more paper to show that the predictions of quantum mechanics couldn't be true. In the paper, he came up with the concept of what's now called quantum entanglement. This notes that under the accepted rules of quantum mechanics, if a particle breaks up into, say, two particles that travel very fast and very far—if one ends up at the far side of the solar system or beyond—an experiment on one can cause an immediate change in certain properties that the other possesses.

In Einstein's mind, the bizarre notion that distant particles could be instantaneously interconnected demonstrated what was "wrong" with the field Bohr, Heisenberg, and the others had begun: clearly, this outrageous implication of their theory meant that the entire thing was unstable. When this didn't persuade the new generation of scientists to change their views, he gave up. There was no use arguing. Although he would continue to criticize quantum theory from time to time, he would never mount a concerted campaign against it again.

EINSTEIN'S OLDER SON, Hans Albert, moved to the United States in 1937. Whatever tension had once existed between them had long since disappeared, and Einstein visited him often in South Carolina, where Hans Albert was working on hydraulic engineering and studying how sediments collect in rivers. They'd stroll in the forests and gossip about Hans Albert's academic research. Einstein was open-minded about it, and when Hans Albert ended up as a professor at Berkeley, he remembered his father still loving to hear about new inventions and clever mathematical puzzles. But if the topic turned to quantum mechanics, Einstein would close down; his views were entirely set.

At one point in the mid-1930s, there had been a chance that Einstein's isolation could end. He'd stayed in touch with the Austrian physicist Erwin Schrödinger, who, although he'd been central to the quantum revolution, remained one of the very few who shared Einstein's doubts about a probabilistic interpretation of quantum mechanics. The two men also shared a certain bohemian attitude toward life. ("It was bad enough to have one wife at Oxford," his biographer remarked about Schrödinger's time guest-lecturing there, "[but] to have two was unspeakable.") They truly liked each other. "You are my closest brother," Einstein wrote him, "and your brain runs so similarly to mine."

Schrödinger, blessedly, had even followed up Einstein's 1935 paper on quantum mechanics with a thought experiment that aimed to show how absurd quantum entanglement (a term the Austrian coined) was. Building on ideas shared in letters with Einstein, Schrödinger proposed his famous scenario in which a cat was trapped in a sealed box with a vial of poison that would be released — or not — depending on whether a decaying radioactive substance within the box let out a single particle. There was a fifty-fifty chance that the cat would die, but the only way to know for sure was to open the box. Until one did that, was the cat alive or dead?

Now known colloquially as "Schrödinger's Cat experiment," this case is used today to illustrate the strange-but-true nature of quantum mechanics. At the time, however, it was held to be a critique of the entire system against which Einstein had railed for so long. In true Einstein fashion, Schrödinger had used his imagination to mount a vigorous attack on quantum theory.

Einstein and Schrödinger were, therefore, predisposed to be partners, and for a time it seemed as if they might have a chance to collaborate more closely. Even though Schrödinger wasn't Jewish, he'd had a tense relationship with the Nazis and had let everyone in the physics community know that he would be happy to be appointed to a post in Princeton, safely across the Atlantic. Had this happened, Schrödinger and Einstein surely would have teamed up. Einstein's thinking on quantum mechanics may well have been clarified, though given his personality it's unlikely that his attitude would have softened as much as Schrödinger's ultimately did. For although quantum mechanics is far from being entirely random — principles such as the uncertainty principle hold very precisely — it is, at its heart, still far from the determinism that Einstein always insisted had to be true.

What Einstein and Schrödinger might have achieved together will never be known, however, for the director of the Institute for Advanced Study, Abraham Flexner, had at this point turned against Einstein — though not for anything having to do with quantum mechanics. Flexner paid Einstein handsomely (not for nothing was it also known as the Institute for Advanced Salaries) but had tried too hard to keep his star attraction under control.

When Einstein had first arrived, Flexner had screened the letters coming to him, and in particular had turned down an invitation for Einstein to visit the White House because he thought it would distract Einstein from his work. For Einstein, that had been infuriating, not just because he hated the idea of being patronized (this was, he wrote in one of his rare curt letters, an "interference . . . that no self-respecting person can tolerate"), but also because the director's meddling was constraining Einstein's affairs in one area of particular importance to him.

Einstein was one of the most active of all émigrés in trying to get refugees away from the growing Nazi power in Europe. He used much of his income to pay for visas for ordinary families; he wrote innumerable letters of recommendation so that ordinary faculty—not just the elite—could get jobs in the United States; he lobbied for changes in policy to allow more of his colleagues to immigrate. The thought that he had been denied a chance to press their case at the highest levels of the U.S. government was intolerable.

When he found out what Flexner had done, Einstein wrote to the president, Franklin D. Roosevelt, and ended up dining at the White House after all. Roosevelt, like many educated Americans of the time, spoke enough German to carry on a conversation in Einstein's native tongue. Along with the European situation, they talked about sailing, which both men loved, and the Einsteins spent the night.

Einstein left the White House having advanced the case of his fellow refugees—but inadvertently having also ruined his last best chance of refurbishing his reputation among his fellow physicists. Flexner was so offended by having his control questioned that—knowing how important Schrödinger could be to Einstein—he blocked any chance of the transfer both scientists so wanted.

Schrödinger ended up in Dublin, where he would remain until the end of Einstein's life.

Even in the isolation of 1930s Dublin—a relatively poor city in a new country that was fiercely separating itself from the United Kingdom—Schrödinger did what Einstein could not. He effectively admitted that he'd given his best arguments and that Bohr and the others had answers for all of them—and so he would accept that his intuition was wrong. He put his old ideas aside and shifted to fresh explorations of the structure of life—work of such insight that it helped trigger the revolution in DNA research that occurred from the 1940s onward.

This was exactly the sort of shift to a new field that had inspired Einstein in the past and might have helped him to revive his career now—if only he had been capable of admitting his error, or at least putting the matter fully out of his mind. But he seemed incapable of doing either. Without Schrödinger's assistance and unable to mount a renewed, more viable offensive against quantum theory, he continued his drift toward the sidelines of science.

Einstein knew he was being shunned. Although the popular press reported on his work with credulous excitement, working physicists scorned it, as in a remark by the acerbic Wolfgang Pauli, writing from Switzerland: "Einstein has once again come out with a public comment on quantum mechanics . . . As is well known, each time he does that, it is a disaster." Another physicist at the Princeton institute remembered that word had gone out to the scientists there that "it would be better not to work with Einstein." The extent of his marginalization became painfully clear when a paper he wrote was turned down by the *Physical Review,* an American journal that was roughly the equivalent of Germany's prestigious *Zeitschrift für Physik.* Einstein was not the sort of man to stand on rank, but this had never happened to him before.

He pretended the failures and rejections didn't matter: "I am generally regarded as sort of a petrified object. I find this role not too distasteful, as it corresponds very well with my temperament." But it was hard to keep up that front entirely, and rather than enduring the humiliation of his own irrelevance, increasingly it seemed that Einstein was simply giving up on the work that others were doing in physics.

Einstein's detachment was plain to see when, in 1939, Bohr himself spent two months at Princeton. Once, the two men had been the closest of intellectual companions ("Not often . . . has a human being caused me such joy"), but this time Einstein almost entirely avoided him: not attending Bohr's talks, not joining Bohr on the long walks the Dane so loved, even avoiding departmental coffees where the one-time friends might meet. When Bohr did go up to Einstein after a seminar, Einstein would speak only in banalities. "Bohr was profoundly unhappy with this," a participant remembered.

But what choice did Einstein have? They were men of the same generation, yet Bohr was still at the center of world research. Einstein was not. Avoiding Bohr meant Einstein could keep his dignity.

By avoiding Bohr, however, Einstein was also taking one more step into isolation—isolation from developments that might have jump-started his own unified field work if he'd been willing to listen to them. Even more tantalizingly, these developments—had Einstein engaged with them—might have led him to make significant contributions to the hunt for the truth in quantum mechanics. But they passed him by, as he did them.

TWENTY

The End

EINSTEIN TRIED TO MAKE a good life for himself outside his beloved field. He sat for sculptors; he made friends with the saintly theologian Martin Buber (discovering a mutual pleasure in Ellery Queen detective stories); he invited the great singer Marian Anderson to stay at his home whenever she was in Princeton. If he was alone, he would improvise for long stretches on the piano. When his cat Tiger was depressed because it had to stay inside during a rainstorm, Einstein's secretary recorded him telling the cat, "I know what's wrong, dear fellow, but I don't know how to turn it off."

Elsa died in 1936, and Mileva Marić—whom he hadn't seen in many years—in 1948, and each loss was a greater blow than he'd expected. Marić's death was especially tragic. She'd been making a decent life for herself in Zurich, supported by Einstein's money; tutoring students in the music and mathematics she'd always loved. But their younger son, Eduard, who'd remained in Switzerland, had been diagnosed with schizophrenia as a young adult. He was in and out of institutions; generally peaceful, and content to dreamily play the piano—one of many similarities to his father that family friends remarked on—but he also experienced manic episodes in which he sometimes became violent. In one of those moments, Marić had been with Eduard and, possibly during a struggle, had collapsed. Three months later, she died in the hospital.

Einstein's sister, Maja, had moved to Princeton just before the war, when her own marriage had fallen apart. (The Winteler family that Einstein had stayed with for his remedial year of high school in Switzerland entered here once again: their daughter Marie had been his first sweetheart, his friend Besso had married another daughter, and Maja had married one of the Winteler sons.) In the hours he spent reading to Maja, Einstein sometimes turned to *Don Quixote,* but more often to Dostoevsky, whose works they both liked, especially *The Brothers Karamazov* and its characters' quest to understand a distant God. Although Ivan, one of the brothers in the book, thought it was impossible to know the Creator ("Such questions are utterly inappropriate for a mind created with an idea of only three dimensions"), Dostoevsky did not, and Einstein was fascinated by the author's conviction.

When Maja died in 1951, Einstein sat on the back porch of his now empty home for hours on end. "I miss her more than can be imagined," he told his stepdaughter Margot when she came out to console him. He continued sitting there, in the hot Princeton summer, at one point gesturing to the sky. "Look into nature," he whispered, almost to himself. "Then you will understand it better." From special relativity he knew that from some perspectives in the universe, the moment of her death hadn't yet occurred. But he also knew that those locations were ones he could never access.

AGE WAS PRESSING IN. In 1952 young players from the Juilliard String Quartet came to visit Einstein in his home, playing for him pieces by Beethoven, Bartók, and one of his favorite composers, Mozart. When he was cajoled to join in, he suggested Mozart's String Quintet in G Minor, and they played together. His hands were stiff and out of practice, but it was a piece he knew well. One of the players remembered, "Einstein hardly referred to the notes on the musical score . . . His coordination, sense of pitch, and concentration were awesome."

Doubts were creeping in, too—darkness seeping into the corners of the great thinker's fabled inner vision. At times he was unsure that his efforts with a unified field theory would work. Once, he wrote that he felt as if he were "in an airship in which one can cruise around in the clouds but cannot see clearly how one can return to reality, i.e. earth." Another time, he admitted to a favorite mathematics assistant that although he could come up with fresh ideas as well as ever, he sometimes worried that his judgment about which were worth pursuing was fading. More often, though, he would tell others with a shrug that he was convinced that in the future, findings in science would catch up with his theoretical work, just as had often happened in the past. Isaac Newton, after all, had disregarded his own qualms about gravity acting instantaneously, and as a result missed out on the breakthrough that Einstein himself had achieved in 1915. The lambda events had shown Einstein the value of holding out for what he was convinced was right. And now, though quantum theory described certain events accurately, he held out hope that it, too, might be just an intermediate step toward a far greater physics that would be discovered in the future.

Early in 1955, his longest-standing friend, the gentle Michele Besso, died. It had been more than a half century before that Einstein had told Marić, "I like him a great deal because of his sharp mind and his simplicity. I also like Anna, and especially their little kid." Now that little boy, Vero, was close to sixty himself. Einstein wrote him and Michele's sister about Michele, explaining how much he had loved and admired him, and adding, "The foundation of our friendship was laid in our student years in Zurich, where we met regularly at musical evenings . . . Later the Patent Office brought us together. The conversations during our mutual way home were of unforgettable charm." This is when he added the remark we've seen before: "Now he has preceded me briefly in his departure from this strange world. This means nothing. For those of us who believe in physics, the distinction between past, present, and future is only an illusion, however tenacious this illusion may be."

By that time, Einstein was seventy-five years old and ill himself, with one of the large arteries from his heart swelling in an aneurysm that doctors explained was going to burst at some unexpected time. There was the possibility of an operation, but medical science was still poor in this area, and there was no guarantee the procedure —if Einstein even survived it—would cure him.

Rather than risk an operation, Einstein decided to continue with his work on the unified field theory, as well as on public statements warning that unfettered nuclear weaponry could destroy all human life on earth. He tried to be stoic. "To think with fear of the end of one's life is pretty general with human beings," he admitted. ". . . The fear is stupid but it cannot be helped." He was anxious about his condition, and no doubt wondered whether science would justify his isolated labors after all.

In early April 1955, Einstein's heart condition worsened. His doctors explained that the aneurysm was tearing. The process would be slow at first, then suddenly speed up. There was more talk of an operation, but Einstein was adamant: "It is tasteless to prolong life artificially. I have done my share." Einstein did ask them what he would experience—how "horrible" the pain would be—but they couldn't tell him anything for sure. An injection of morphine helped slightly.

By Friday, April 15, he was in such pain that he was driven to Princeton Hospital. When his stepdaughter Margot came, she barely recognized him: pale, his face twisted in pain. Even so, "his personality was the same as ever," she remembered. "He joked with me . . . and awaited his end as an imminent natural phenomenon." His older son flew in from Berkeley, where he now worked as an engineering professor. Speaking to Hans Albert, Einstein gestured to his equations—which were yet another effort at creating a unified field theory to bring together all known forces in a clear, predictable manner. He said, wryly: "If only I had more mathematics."

Soon he felt a little better, and even asked for his glasses, as well as a pencil and his papers, to work on those calculations some more. But then in the early morning of Monday, April 18, the aneurysm burst.

He was alone, bleeding to death very quickly. He called out to a nurse, and when she arrived, he whispered to her. But she spoke no German, and so had no idea what the old man said before he died.

Epilogue

ONE DAY AROUND 1904, when Michele Besso's son, Vero, was young, a friend of his father's made the boy a splendid kite. The three then walked into the countryside, in the direction of a small mountain south of Bern, taking the kite with them. At the foot of the mountain, one of the adults launched the kite and, once it was airborne, put the string into the boy's hand.

Vero would remember this family friend clearly in later years, for the man "was always in a good mood, he was amusing and jolly, and above all he knew lots of things." In particular, Vero never forgot how, on that day, as the kite soared through the air, the man who had made it—Mr. Einstein—could explain to him *how* it flew.

Einstein was a man of insatiable curiosity and great kindness. Like anyone, he had his faults, and over the course of his life they were magnified by his outsize achievements. But his underlying impulse was pure. If the end of his career was tragic, it was only because he became locked into mistaken lessons from his past.

He had dreamed of being redeemed by history when it came to quantum mechanics, but the very opposite has taken place. In the 1950s and 1960s, researchers developed ways to test Einstein's belief that quantum mechanics was just a temporary step to a more certain future theory, one that would dispose of the randomness he loathed and provide a more logical, orderly explanation of how the universe functions. When those tests were carried out in the 1980s, however, they confirmed that Heisenberg, Bohr, and the others had been right: the uncertainty principle is rock solid. The world does not operate in the determinist way Einstein wanted to believe it did.

The only thing certain, at least at the atomic and subatomic levels, is a certain degree of randomness.

In time, some of Einstein's own efforts to disprove quantum theory would be turned against him. Even the paper he coauthored in 1935, showing that quantum mechanics would allow distant particles to be "miraculously" entangled, has only strengthened the now accepted view. Those entangled particles have actually been created and are being used in the first generations of quantum computers being built today, in the twenty-first century.

In a vast range of important fields, however, Einstein's approach and findings have been so completely accepted that generally they are not even acknowledged as coming from him: they just "are." Our fundamental understanding of photons, of lasers, of low-temperature physics, and, of course, of relativity stems directly from his papers written in Bern, Zurich, and Berlin. Collectively, these achievements are rivaled only by those of Newton in their impact on our lives and in the way they deepened our understanding of the cosmos.

Although Einstein's particular approach to finding a unified field theory failed, many researchers in subsequent generations have been inspired by the knowledge that the world's greatest mind spent so many years on this hunt. Einstein's fruitless quest is, for example, what helped inspire physicist Steven Weinberg and others in their successful work unifying electromagnetism with the weak force that operates inside the atom, an achievement for which they were awarded the Nobel Prize.

With general relativity, which has been symbolized as G=T in this book, Einstein's work is linked to some of the most stunning discoveries of modern times. His insights about gravitational lensing do show that when we look at distant clusters of galaxies, we should be able to see at least something of what's behind them. That swerving of light is just what Eddington measured in his 1919 photographs of light bending near the sun.

The more mass there is in those clusters, the more the space around them will bend, and the more powerful that distant lens will be. This today helps to allow us to estimate how much mass there is in such galactic clusters—in informal words, to "weigh" them. The results have been startling, and this has helped show that what we thought filled the universe—stars and planets and the like—is only a small part of the full mass those clusters contain. Most of what exists in the universe is entirely invisible to us, and what it is composed of we do not know. This unseen "stuff" that Einstein's work allowed us to discover is called dark matter and is a major topic of research.

Not everything about G=T has worked out so neatly, however. When it comes to the lambda, the greatest of ironies has arisen. Einstein had been reluctant to put that extra term into his great 1915 equation, despite its effectiveness in providing a repulsion that pushed outward against gravity. He was delighted when Hubble and Humason in 1929 seemed to show that the universe was expanding at a steady rate, and so no such lambda term was needed. But starting in the 1990s, new findings began to suggest that, inadvertently, Einstein might have been right after all when he brought in the lambda. The universe is not only spreading outward, but something is propelling it apart at an ever faster rate. That enormous force of repulsion has been labeled dark energy and is exactly what a revised lambda term could account for. If this holds up, it would mean that to some extent, what Einstein had felt as his mistake wasn't, in fact, an error—and all the stubbornness that came from it was unnecessary. Research on a new, revamped cosmological constant is now of great interest, because of its implications for Einstein's work and its connection to burgeoning new subfields of physics.

These are humbling findings. Everything we see and thought we knew—all the continents and oceans on earth; all the planets and stars beyond—makes up only a very small part of the universe. Dark matter makes up perhaps 25 percent of all there is; dark energy perhaps 70 percent. The entire world we know is but a small,

5 percent fragment, surfing on an unseen immensity. The dark energy component is what necessitates a lambda term after all, modifying Einstein's great 1915 work. The dark matter is different, and to a great extent can be seen as just another bit of "mass" to plug into his otherwise still valid equations.

Einstein himself pondered how much of the universe's vastness can be perceived by the human mind. In 1914 he had written to his friend Heinrich Zangger, "Nature is only showing us the tail of the lion. But I have no doubt that the lion belongs to it, even though, because of its colossal size, it cannot directly reveal itself to the beholder." The underlying truth is hard to perceive. But perhaps one day another genius like Einstein—avoiding the mistake of hubris this time—will show us the behemoth itself.

Acknowledgments

WRITING THE FIRST draft of this book, I felt as if the Muses were dictating the story, but my friends—a cynical lot—thought that if the Muses *were* providing dictation, it was curious that they included so many clumsy phrases and dull ~~repeated~~ repetitions that had to be fixed, and Shanda Bahles, Richard Cohen, Tim Harford, Richard Pelletier, Gabrielle Walker, Patrick Walsh, and Andrew Wright jumped in with somewhat disturbing eagerness to get that done.

While they were ~~savaging~~ gently improving the manuscript, Michael Hirschl prepared deft illustrations for the main text, and Mark Noad did the same for the online appendix. At one moment, when a large chunk of edited text disappeared into the ether, Carrie Plitt most wonderfully managed to re-create it; Yuri from the Regent Street Apple Store in London helped when the ether called again. Arthur Miller and James Scargill saved me from a number of errors (though neither is responsible for any errors I may have subsequently added). In New York, Alexander Littlefield read the entire manuscript and as deadlines neared made a tremendous number of improvements with a calmness that was inspiring to behold. After that, the rest of his team joined in, and it was a pleasure to receive their help: Beth Burleigh Fuller, Naomi Gibbs, Lori Glazer, Martha Kennedy, Stephanie Kim, Ayesha Mirza, and—in distant New Hampshire—Barbara Jatkola, who copyedited the whole thing. In London, Tim Whiting provided excellent counsel and support, and thanks are also due to Iain Hunt, Linda Silverman, Jack Smyth, and Poppy Stimpson.

A generation of Oxford students helped in my initial thoughts on this project by sitting through the Intellectual Tool-Kit lectures in which I first tried out several of the ideas presented here. Going back to the mid-1970s, I had the honor of studying under Chandrasekhar, who worked with many of the principals in this story. (He was the young guest sitting with Rutherford and Eddington at the start of interlude 2.) From the late 1970s, I fondly remember a long afternoon with Louis de Broglie in Paris, his memory of the days when quantum mechanics was being created still utterly clear.

None of that would have led to this book, if it had not been for the remarkable fact that after many years of being a single dad I met Claire. When I proposed marriage—after waiting an interminable eight days from when we first met—she put one finger to my lips, and then whispered: Of course. I had no idea life could offer this.

I'd been too daunted to attempt this book before, but with the confidence that gave me I was able to move forward. Once I began, Mark Hurst deftly showed me how to focus the story, while Floyd Woodrow, the most inspiring of men, showed me how to remain focused.

As I was working through the chapters, my children Sam and Sophie received updates at weekly, sometimes daily, and once (sorry, guys) even hourly intervals detailing my progress. Their confidence that this story needed to be told was the most invigorating of motivations.

I dedicated it to Sam because when he was young, and important matters—birthday presents or new computer games—existed too far in the future for mortal souls to wait, I'd explain that if we could simply get into an Einstein rocket, we could reach those future dates in just a few minutes of our time. I loved the way he trusted this. If people do manage to create such devices to accelerate us through time, it will be members of his generation, not mine. And if that generation avoids the hubris that brought Einstein down, I will be delighted.

Appendix

A Layman's Guide to Relativity

This book stands on its own, but in this appendix you can delve a little more into how relativity works. Readers who skip it won't find their appreciation of the book affected. Readers who are true gluttons will find a 22,000-word download at davidbodanis.com taking them further.

WHY TIME CURVES: THE CASE OF KING KONG

The idea that it's not just space that gets curved, but time as well, was first properly developed by one of Einstein's old professors, Hermann Minkowski, at a lecture in Cologne, Germany, in 1908. He had been thinking about Einstein's 1905 work and noticed that "Einstein's presentation [of special relativity] is mathematically awkward—I can say that because he got his mathematical education in Zurich from me."

To extend Einstein's work, Minkowski began by laying out an image where space was envisioned as a horizontal plane, and time as a vertical axis sticking up from it. One can think of this as a large table with a spindle or candlestick rising up from it in the middle. Everyone was used to seeing the two realms as separate, but that was what Minkowski wanted to change. It made sense to him, for "the objects of our perception invariably include places and times in combination. Nobody has ever noticed a place except at a time, or a time except at a place."

It would be better, Minkowski declared, not to speak of ordinary locations plus a separate time, but rather of a single unitary thing called an "event." To describe the mixed "space-time" that all possible events fit within, one would just need to make lists of four numbers.

That sounds abstract, but it's something we do all the time. Suppose your great-grandfather was strolling in New York one crisp evening in the spring of 1933 and saw a large, hirsute creature on the top of the Empire State Building, 1,400 feet up. He wants to alert the press. Once he finds a telephone and rings the *New York Herald Tribune,* he could say, "It's—it's—on the top of the Empire State Building, and, oh God, I see it right now!" But if he and the *Herald Tribune* reporter both understood Minkowski's symbolic shorthand, he could more swiftly say, "5th Avenue, 33rd Street, 1,400 feet, 8:30 p.m.!" If they both understood Manhattan's grid system, he could be even quicker, simply saying, "5, 33, 1,400, 8:30!" The paper's photographers would know exactly where to head: the corner of 5th Avenue and 33rd Street, way up on the 1,400-foot-high spire, where at least at 8:30 p.m., the city's largest resident was to be found.

But imagine King Kong is publicity shy and tosses a zip line across central Manhattan to the top of the shinily attractive Chrysler Building. Blond actress in hand, he starts gliding across to that safer refuge. If your great-grandfather were still watching and had the phone, he could tell his *Trib* contact the new changing coordinates. Five seconds into the gliding journey, he might call out "5, 35, 1,380, 8:30:05"; five seconds later, it might be "5, 36, 1,340, 8:30:10"; and so on. The figures would click along till the pair arrived atop the slightly shorter Chrysler Building on 42nd Street.

That's what Minkowski meant when he said that every distinct event—every distinct location in space and time—can be identified by a grouping of four numbers. Listing every possible event in the universe would produce an enormous book, one in which all the settings for past or future history are written out. In his 1908 lecture, Minkowski joked about how presumptuous this task was: "With this most valiant piece of chalk I might project [all those locations] upon the blackboard." It's precisely what many religions imagine their God to be able to do. But that didn't stop him from observing that this is how it could be done.

Now for the big question. Does what happens to the first three numbers—the ones describing location in space—have to be linked with the fourth number, the one describing location in time? If it does, then space will not be separate from time, but each will have to be brought in to fully locate what is happening.

To answer that, Minkowski looked at how to work out distances between any two events. For your great-grandfather, the distance between the starting event, where Kong is atop the Empire State Building, and the second event, where Kong has landed atop the 300-foot-shorter Chrysler Building, would be something like "3 avenues, 8 streets, 300 feet, 2 minutes." But remember that time passes at a different rate for objects that are moving relative to you. That is key. Kong isn't going especially fast past your great-grandfather as he glides over Manhattan, but he's going to experience a travel time that's just a very slight amount less than the full two minutes your great-grandfather sees.

(Why does time vary like that? Well, suppose you're watching a friend bounce a ball up and down in a stationary car near you. Clearly you and she will agree on how far it's traveled. Now, however, have her start driving the car while you remain by the roadside watching. She'll see the ball she's bouncing continue to go straight up and down beside her. You'll see it take a longer path as the car moves forward.

(Now suppose it's not a ball she's bouncing, but a light beam. You'll each see it travel at the same speed (for as Einstein made clear, that's how light works). This is what's odd. She'll see the light beam in her car cover a short distance. You'll see it moving at the same speed—for that's the only rate light can travel—and cover a longer distance.

(How can something cover two different distances while traveling at the same speed? The only answer, Einstein realized, is if you see time in the moving car slow down, so that there's more time for the light beam to travel the longer distance. Any object moving relative to you will experience this, be that cars, rocket ships, or even imagined fast-gliding apes.)

The effect stands out if we imagine faster travel. What if Kong doesn't stay on the Chrysler Building but, worried about the *Herald Trib* press cars scurrying close by, at 8:32 p.m. he and his date jump into a rocket ship and circle around the galaxy until they land back on top of the Chrysler Building on what you measure as February 8, 2017. You hurry down there, push through the massed photographers and the producers shouting out reality TV offers, and huddle with the great beast and his actress friend. You ask if they'll help you with the Minkowskian calculation to determine the space-time distances they've crossed.

They nod yes and show you the log in which they've carefully kept note of their travels. You read it and then look up, puzzled. To you it's obvious what the "distance" is between the event when Kong was last seen atop the Chrysler Building and the situation today. That event took place at 8:32 p.m. on March 2, 1933, and now you're standing at the same place, so the difference is "0 avenues, 0 streets, 0 height, 83.9 years." But Kong's log shows a much briefer amount of time, due to the distortions of time that took place over the immense distances he traveled in his high-speed journey.

The point is a profound one. Different individuals constantly enter onto separate "tracks" of time. It's not just you and our imagined far-traveling Kong who won't agree when it comes to measuring the distance between two events. All of us travel at different rates, and—if one looks closely enough—we will all have these disagreements about what has elapsed between two events.

FINDING YOUR WAY THROUGH TIME AND SPACE

This seems like a recipe for chaos, as if we're living in a universe that's wildly unconnected, with each of us in separate worlds, crashing against one another with no rhyme or reason. But Minkowski showed that although space and time don't fit together by the simple sort of subtractions between events described in the previous section, they do fit together another way. There is a new kind of distance between any two events—what he termed the "interval" —which everyone will agree on, however they've been moving. Although your space and your time might be different from mine, Minkowski found that the curious number x^2-c^2t^2 will always bring us to the same result. (Here *c* is the speed of light, *t* is the difference between the time entries in the two events, and *x* is an elapsed distance between all the space entries—it's a matter I elaborate on my website. Since x^2-c^2t^2 describes a hyperbola, a number of geometrical diagrams fit there to help.)

Einstein resisted this blending at first, calling Minkowski's work *überflüssige Gelehrsamkeit*—"superfluous erudition"—but he soon came around and made Minkowski's solution integral to his own later work in relativity. It's a fabulous solution. We no longer have to think of our universe as an ungainly pileup, with three-dimensional space here, and a one-dimensional time sticking out to the side of it at a right angle somewhere else, and everybody whirring along like Magritte characters on their own isolated airport walkways. Instead, we live in a combined "thing" called space-time.

The interval—that strange "distance" x^2-c^2t^2—is at the heart of the trade-off that occurs between space and time. In ordinary space, distances add up, and then, separately, time adds up as well. In space-time, because the components are linked in this particular way (with the time we see of another individual slowing down as his velocity relative to us speeds up), that doesn't happen. Rather, it's as if movement in space-time works through two odometers, one of which is constantly subtracting from the other.

The idea of mixing time with space can sound mystical, but think of looking at a circular watch. Viewed face-on, it seems to have an equal amount of "horizontalness" and "verticalness." Tilt it slightly, however, so you're viewing it at an angle, and you won't see a perfect circle anymore, but an ellipse. Some of the verticalness seems to have vanished.

We're not bothered by this, for we know that the verticalness would still be there if we took a complete measurement of the watch. Its vanishing is simply an artifact of our restricted position in viewing it. The spatial dimensions we're used to are similar. We know that on earth we can walk due east or due north, but we can also do a bit of each at the same time—that is, we can stroll northeast. The directions north and east might seem distinct, but they fit within a greater unity we can perceive. Similarly for the way that fans in a basketball arena see the hoop foreshortened in different ways. Those whose seats put them precisely ten feet up see it as a horizontal line; those in other positions might see it as an ellipse. But that doesn't mean they think those distortions are the truth. Once they stand up and walk around, they know they can see the hoop from enough angles to get the full picture.

In the four-dimensional space-time that Minkowski showed we live in, however, it's impossible for us fragile, carbon-based organisms to step back and see the full arrangement—to see all of space and all of time at once. Yet with his abstract symbols, we can tell it's there, with all the parts—all space and all time—inextricably linked.

THE EQUATION OF THE UNIVERSE

How did Einstein bring all that together? His 1915 equation looks so different from what's taught in high school or even basic university math classes that most people on first glance think nothing about it can be understood. Even in its most condensed modern form, the equation comes out as the not especially inviting $G_{\mu\nu}=8\pi T_{\mu\nu}$. But once we realize that much of that is simply a deft shorthand for listing various mixes of things, it starts to become clearer.

To understand that shorthand, imagine going back to one of the restaurant-cafés that Einstein used to like sitting in when he was a university student. Suppose the menu is extremely small, just schnitzel and beer, and to save time the waiters don't write down in full their customers' orders, but instead use little grids printed on their order pads:

If a waiter sends in the order

1 0
0 1

the chef knows to supply a double order of schnitzel (because the first "1" is in the slot where the two schnitzel labels intersect), a double order of beer, and nothing else.

If the chef decides to go wild and offer a third menu item—roasted potatoes!—the restaurant will need to print up fresh pads with a slightly larger grid. The best form of roasted potatoes in Switzerland is called *rösti,* and so the new grid will read like this:

If a waiter now writes

0 0 1
0 3 0
0 0 0

the chef knows to produce one tray of mixed schnitzel and rösti, and one tray with three double orders of beer. It's exceptionally unhealthy, remarkably tasty—and very efficiently summarized by that simple pattern of numbers.

Suppose there are dozens of such restaurants in Zurich, and they decide to stop competing with one another. Instead of offering a choice of how much of any mix can be ordered, each restaurant sticks to a particular quantity for every meal it serves. In one restaurant, *every* customer gets a mixed tray of schnitzel and rösti, and three trays with double orders of beer—and that's it. Another Zurich restaurant has a sign outside the door with a blowup of the waiters' order pad filled in like this:
The grid reads

1 0 0
0 0 0
0 0 1

so everyone knows that this restaurant's offerings are double schnitzels and double rösti, and never anything else. The other restaurants lock in to yet other possibilities. Each simple arrangement of numbers lets you know what culinary treats you will henceforth eat if you enter a restaurant and remain inside forever.

Now back to relativity. Suppose we're ordering not food, but the shape of a universe. First of all, we need to know what the constituent dimensions are—the equivalents of the schnitzel, beer, and rösti. In the case of a two-dimensional flat surface, like the one our postage-stamp-shaped Mr. A. Square lived on, those constituent parts are the changes in distance in the east-west direction, *dx*, and the changes in distance in the north-south direction, *dy*.

That sets up the waiters' order pad, and to flesh it out we now need to know which particular permutations of those parts will be on offer. Once we have those two sorts of data—what range of possibilities are on offer to set up our grid, and then what particular choices are made from that range to fill in our grid—we know a great deal about the world we are going to enter.

That type of "mixed grid plus entries" is very close to what is called a metric tensor. The name is illuminating. Measures of distance using the Greek root *metron*, or "meter," were popularized with the new French measuring system introduced in the eighteenth century: the metric system. Metrics are simply a way of establishing how things fit together. The cluster of numbers that serves as an order in each of our hyperefficient Zurich restaurants defines how the constituent food components fit together. It's the restaurant's metric—*its* way of organizing things. In our physical universe, the cluster of numbers that serves as an "order" defines how our universe's component dimensions fit together.

CREATING OUR WORLD

In Mr. A. Square's Flatland world, the background grid will allow for different mixes of *dx* and *dy:* different amounts of east-westness or north-southness. That means the empty grid will look like this:

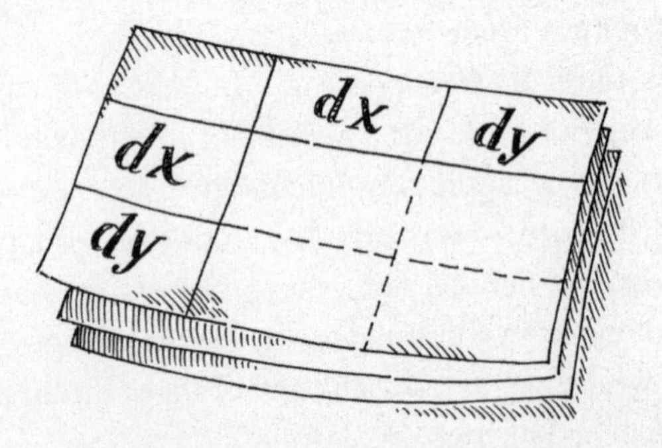

How to fill it in? We know that the very definition of a flat surface is that the Pythagorean theorem holds: that if there's a right triangle with sides *dx* and *dy*, and a hypotenuse *ds*, then the parts click together so that $dx^2+dy^2=ds^2$. We can summarize that in the Flatland order form by filling it in as shown:

There are double orders of *dx* and double orders of *dy*, and no odd mixes of them. Everything is neat: right triangles fit together, squares don't bulge, and it's fair to think that time will similarly be at "right angles" to space. Stick to combining the parts in this most straightforward way, and you'll create the Flatland world. In the Bible's book of Job, God asks, "Where wast thou when I laid the foundations of the earth? . . . Who hath laid the measures thereof . . . or who laid the corner stone?" This is the closest secular equivalent to how that can be done.

What's great about this approach is that it can easily be expanded. A benevolent deity looks down on his dominion, orders up more constituents, and lo, a restaurant—a universe!—is created with a much-enlarged order form.

Einstein's equations were built out of analogous grids but allow very different worlds. They were bigger than Flatland's, of course, for instead of having 2-by-2 boxes, allowing for just two dimensions of space (east-west and north-south), they had 4-by-4 boxes, so that three dimensions of space and one dimension of time could be combined. Also, they weren't usually going to be filled in with such simple instructions as that of the Flatland grid, where the sequence of 1s on the diagonal meant the only mixes were crisp, neat, straight ones. That would have been like landing in a restaurant world where separate menu items were never mixed: what in four dimensions—i.e., with four ranges of possible items to be ordered—would look like this:

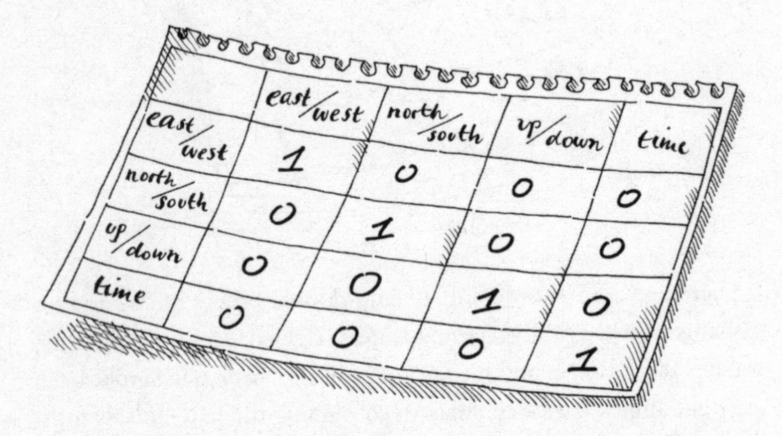

That's a dull, flat world—like the one on page 74 (left).

It's more exciting—and, Einstein realized, more realistic—to allow for mixes of the different dimensions to take place. That way, for example, a bit of east-westness could be ordered in combination with a bit of up-downness. This is like one of the restaurants where the chefs ventured away from the standard offering of double orders of beer or double schnitzels. Instead, moving off the diagonal line, they offer variable mixes of food. That's closer to the universe we exist in. The result is like the one we saw on page 74 (right), where at some places up-downness is mixed with east-westness, and other mixes occur as well.

Words get unwieldy when you use them to fully describe what

goes in all the many possible slots. Instead of saying, for instance, "Here's the value that will go in the box where the third column intersects the fourth row," it's quicker just to say "Here's the value for box_{34}." Einstein and Grossmann went further, and instead of the word "box," they began to use the letter *g*. Then, when they wanted to show that their boxes would work through a full range of subscripts, they switched from numbers to Greek subscripts. When Einstein wrote g_{34}, he was referring to the value that would go in the box at the intersection of the third column and the fourth row; when he wrote $g_{\mu\nu}$, he was referring to the whole range of sixteen boxes in his menu grid. It's much like the way we refer to the boxes in spreadsheets today.

Thus the equation Einstein came up with in 1915, what in its neatened form is $G_{\mu\nu}=8\pi T_{\mu\nu}$. The capital *G* on the left side is still a bit more complicated than what can be described here, but it has at its heart the expression $g_{\mu\nu}$: a set of values that fit within a 4-by-4 grid and lay out how space and time at a particular location are ordered. The $T_{\mu\nu}$ is similar and has at its heart another 4-by-4 grid whose values describe what's *in* that location in space-time: the mixes of energy and momentum we'd find there.

Most of all, magnificently, Einstein recognized the deep connection between the two sides. One didn't have to go to every possible location in space-time and then measure all the mass and energy there to fill in the two sides. That would be—and it is difficult to emphasize how much of an understatement this is—a lengthy task to complete. Instead, through Einstein's genius, half the work is already done for us. Identify some particular arrangement of space and time on the left, and you will have a great start in knowing how mass and energy are operating there. Or you can start on the right, measuring what's in the *T* grids, and then by the magic of his equation you will immediately be able to travel to the left side and start describing the geometrical configuration of space and time there. And of course if the values on the left side are so great that they would pull the entire universe into collapse, you can subtract some fraction of that left side so that everything balances without collapsing—and that is what Einstein did with his additional lambda term in 1917.

Solving this relation is hard, for the entries in the boxes on either side don't just sit still. They vary from different perspectives. For example, if I see an object as stationary, you who are moving relative to me will *not* see it as stationary. But since it's moving for you, it now has kinetic energy for you, and by the equivalence of energy and mass—remember $E=mc^2$—you'll genuinely experience it having greater gravitational attraction than I would.

Similarly, a mass that has one length when seen by a relatively stationary observer will be contracting when seen by a moving observer. But since the mass doesn't change, while the volume does, its density will be higher, and that, too, will have to be incorporated. How to write such matters out so our differing perspectives each hold true? That's what took Einstein and Grossmann so long.

Luckily, there are ways to make the calculations a little bit easier. For example, each side of Einstein's relation has deep symmetries around the diagonal axis that goes from the top left to the bottom right, because everything on one side of that line has a duplicate in a matching position on the other (just the way an order of schnitzel plus beer yields the same result as an order of beer plus schnitzel). This means that instead of there being sixteen independent slots, and so sixteen separate equations, there are only the four along the middle diagonal, and then the six others above it. This makes $G=T$ a mere ten tangled equations—which is, small mercies, at least easier to deal with than sixteen.

WHAT EINSTEIN SAW

It's often possible to get rich insights without having to solve the equations at all. To understand how time bending depends on gravity, for example, imagine the space entrepreneur Elon Musk wants to check out one of his rocket ships before launch. He climbs into the bottom of the rocket ship, looks at the watch on his wrist, and then peers up to the top of his rocket—the inside having been hollowed out enough that he can see all the way—where another clock is waiting. He can tell the two clocks are synchronized because flashes of light are coming down from the top one and arriving at what his wristwatch shows are rock-steady intervals.

Everything seems fine.

But then suddenly, his good friend Jeff Bezos, watching from outside, punches a red button. Musk feels that his rocket is shooting away from the earth. He's flung back against the bottom, and—delighted that Bezos has given him this opportunity to experience general relativistic effects—notes something curious. Somehow the light flashes from the top, or front, are arriving more quickly now than they did before. He's puzzled by this. He knows that the length of his rocket hasn't changed. Nor has the speed of light.

So why are these flashes reaching him more quickly?

He puzzles some more and then realizes what is happening. Because he's accelerating, the back of the rocket, where he is, is approaching where the front had been ever more quickly than it had before. (That's what acceleration means, as opposed to constant velocity.) Musk, at the back, is intercepting the light flash from the front before it's gone the full length of his rocket ship: before, that is, his wristwatch has had a chance to mark a full second.

He can draw only one conclusion. The light flash from the front is arriving too "soon." It used to send a flash every second by his wristwatch's time. Now it's sending a flash sooner than that. The clock must be running fast.

If this only happened inside rocket ships, the effect might be considered as due to the rumbling vibrations of the engines. But remember Einstein's insistence that if there are no windows, the passenger can't know for sure if he has been flying away from the earth. Maybe he's been tricked and is still down on the ground, and it's gravity that is holding him to the floor. (This is, as we've noted, like being pressed back in an accelerating sports car. If your eyes are closed and there's no shaking, it will feel just the same as if you're being pulled backward by a huge gravitational source behind you.)

Since no observer in these circumstances can tell whether he's on the ground or flying away from the earth, this means that the different rates of time—the different speeds that the clocks tick—will occur in a gravitational field as much as in an accelerating vehicle. Identical clocks show time passing more quickly "high up," where gravity is weaker. Time passes more slowly "low down," where gravity is stronger.

This sounds preposterous, but it's true. When GPS satellites whiz overhead, then in accord with special relativity their high velocity contributes to making time on board slow down. But since satellites also orbit at 12,500 miles up, where gravity is several times weaker than on the ground, the effect that Musk's imagined experience demonstrates also comes into play. From this second effect, in accord with general relativity, time on the satellites flows more quickly than it does for us on earth, where our denser gravitational field slows time down.

Which factor dominates? In the case of our GPS satellites, the speeding of time due to the lesser gravity in their high orbits adds 45,000 nanoseconds of extra time to their existence each day, while the slowing of time due to their great speed takes away only about 7,000 nanoseconds each day. The net difference is a 38,000 nanosecond gain. That's the figure engineers use to "reset" our GPS systems each day, to keep time in our world in synch with time on the satellites. Without this correction, we'd soon end up miles off course.

There's more. The greater the difference between gravity at two places, the greater the general relativistic effect. Time just above the surface of the sun proceeds a minute slower each year than it does on earth. Time very near a black hole proceeds many millions of times slower. We would see (subject to certain dimming effects) an astronaut who is falling into the black hole seem to move in incredibly slow motion, while to him time is normal, and it's the galaxy outside that has sped up, with life speeding along at millions of times its usual rate. In his last moment, he could, in theory, see entire civilizations rise and fall.

It would be hard for him to actually observe such sights, however, and not just because of the limitations of any telescope he's brought along. A gravitational gradient strong enough to produce time rates that are so different will also produce very different pulls on his body. His raised hand will receive a certain amount of gravitational pull; his foot, if it's closer to the hole, will receive a greater—a VERY much greater—pull. Other effects are going on, but this alone is enough to create what's called "spaghettification," where even the strongest material is pulled apart. However his investments back on earth might have turned out, he would soon not be in any condition to enjoy them.

Credits

Illustrations on pages 60, 65, 66, 74, 76, 87, 89, 160, 244, 245, 246, 247, and 248 are by Michael Hirschl, © 2016.

page

x: Esther Bubley, The LIFE Images Collection / Getty Images
2: SPL / Science Source®
10: Besso Family, American Institute of Physics, Emilio Segrè Visual Archives
11: American Institute of Physics, Emilio Segrè Visual Archives / Science Source®
12: ullstein bild / Pictures from History
42: ullstein bild / AKG
45: From *Flatland: A Romance of Many Dimensions,* Edwin Abbott Abbott, 1884
46: From *Flatland: A Romance of Many Dimensions,* Edwin Abbott Abbott, 1884
82: Keystone-France / Getty Images
96: © UPPA / Photoshot
114: © Jiri Rezac
124: Sergey Konenkov, Sygma / Corbis
129: RIA Novosti / Science Source®
138: Harvard College Observatory / Science Source®
144: AP Images
149: Margaret Bourke-White, Time Life Pictures / Getty Images
151: SPL / Science Source®
155: Mondadori Portfolio / Getty Images
170: Albert Einstein, Courtesy of the University of New Hampshire
173: ullstein bild / Getty Images
179: ullstein bild / Rainer Binder
187: Francis Simon, American Institute of Physics, Emilio Segrè Visual Archives / Science Source®
194: Science & Society Picture Library / Getty Images
202: American Institute of Physics, Emilio Segrè Visual Archives / Science Source®
210: Bettmann / Getty Images

Bibliography

Here are a handful of especially worthwhile books, mostly for the general reader. In each section, I've marked two of my favorites with asterisks. There's also a long annotated version of this list at davidbodanis.com. In addition to these is the fundamental *The Collected Papers of Albert Einstein* (Princeton: Princeton University Press, 1987–), now at fourteen volumes and counting.

LETTERS, ESSAYS, AND QUOTATIONS

Albert Einstein–Michele Besso Correspondance, 1903–1955. Translated and edited by Pierre Speziali. Paris: Hermann, 1972.

*Born, Max. *The Born-Einstein Letters, 1916–1955: Friendship, Politics and Physics in Uncertain Times*. Translated by Irene Born. London: Macmillan, 2005. First published 1971.

Calaprice, Alice, ed. *The Ultimate Quotable Einstein*. Princeton: Princeton University Press, 2011.

*Einstein, Albert. *Ideas and Opinions*. London: Folio Society, 2010.

Solovine, Maurice. *Albert Einstein: Letters to Solovine*. New York: Philosophical Library, 1987.

BIOGRAPHIES (AUTHORS WHO KNEW HIM)

*Frank, Philipp. *Einstein: His Life and Times*. New York: Da Capo Press, 2002). First published 1947.

*Hoffmann, Banesh. *Albert Einstein: Creator and Rebel*. New York: Viking, 1972.

Pais, Abraham. *Subtle Is the Lord: The Science and Life of Albert Einstein*. New York: Oxford University Press, 1982.

Seelig, Carl. *Albert Einstein: A Documentary Biography*. London: Staples Press, 1956.

BIOGRAPHIES (MORE RECENT)

Folsing, Albrecht. *Albert Einstein: A Biography.* Translated and abridged by Ewald Osers. New York: Viking, 1997.

*Isaacson, Walter. *Einstein: His Life and Universe.* New York: Simon & Schuster, 2007.

*Neffe, Jurgen. *Einstein: A Biography.* Translated by Shelley Frisch. New York: Farrar, Straus and Giroux, 2007.

Renn, Jürgen. *Albert Einstein: Chief Engineer of the Universe.* Hoboken, N.J.: Wiley, 2006.

REFLECTIONS, AND SPECIAL TOPICS

French, A. P., ed. *Einstein: A Centenary Volume.* Cambridge, Mass.: Harvard University Press, 1979.

Galison, Peter. *Einstein's Clocks, Poincaré's Maps.* New York: Norton, 2003.

Gutfreund, Hanoch, and Jürgen Renn. *The Road to Relativity: The History and Meaning of Einstein's "The Foundation of General Relativity."* Princeton: Princeton University Press, 2015.

Holton, Gerald, and Yehuda Elkana, eds. *Albert Einstein: Historical and Cultural Perspectives.* Mineola, N.Y.: Dover, 1997. First published 1982.

*Levenson, Thomas. *Einstein in Berlin.* New York: Bantam Books, 2003.

Miller, Arthur I. *Einstein, Picasso: Space, Time, and the Beauty That Causes Havoc.* New York: Basic Books, 2001.

*Schilpp, Paul Arthur. *Albert Einstein: Philosopher-Scientist.* LaSalle, Ill.: Open Court Press, 1949.

Stachel, John. *Einstein from "B" to "Z."* Boston: Birkhäuser, 2002.

Stern, Fritz. *Einstein's German World.* Princeton: Princeton University Press, 1999.

RELATIVITY IN PARTICULAR

Einstein, Albert. *Relativity: The Special and the General Theory (A Popular Account).* Translated by Robert W. Lawson. New York: Random House, 1995. First published 1916.

Ferreira, Pedro G. *The Perfect Theory: A Century of Geniuses and the Battle over General Relativity.* New York: Houghton Mifflin Harcourt, 2014.

*Geroch, Robert. *General Relativity, from A to B.* Chicago: University of Chicago Press, 1978.

*Susskind, Leonard. General Relativity. Online course. The Theoretical Minimum, Stanford Continuing Studies, http://theoreticalminimum.com/courses/general-relativity/2012/fall.

Taylor, Edwin, and J. Archibald Wheeler. *Spacetime Physics: Introduction to Special Relativity.* New York: W. H. Freeman, 1992.

Thorne, Kip. *Black Holes and Time Warps: Einstein's Outrageous Legacy.* New York: Norton, 1995.

Wald, Robert M. *Space, Time, and Gravity: The Theory of the Big Bang and Black Holes.* Chicago: University of Chicago Press, 1992.

Will, Clifford M. *Was Einstein Right?: Putting General Relativity to the Test.* Oxford: Oxford University Press, 1993.

QUANTUM MECHANICS

Fine, Arthur. *The Shaky Game: Einstein, Realism, and the Quantum Theory.* Chicago: University of Chicago Press, 1996.

Kuhn, Thomas S. *Black-Body Theory and the Quantum Discontinuity, 1894–1912.* Chicago: University of Chicago Press, 1978.

*McCormmach, Russell. *Night Thoughts of a Classical Physicist.* Cambridge, Mass.: Harvard University Press, 1982.

Polkinghorne, John. *Quantum Theory: A Very Short Introduction.* New York: Oxford University Press, 2002.

*Stone, A. Douglas. *Einstein and the Quantum: The Quest of the Valiant Swabian.* Princeton: Princeton University Press, 2013.

OTHER PLAYERS

*Cassidy, David. *Uncertainty: The Life and Science of Werner Heisenberg.* New York: W. H. Freeman, 1992.

Halpern, Paul. *Einstein's Dice and Schrödinger's Cat: How Two Great Minds Battled Quantum Randomness to Create a Unified Theory of Physics.* New York: Basic Books, 2015.

Heilbron, John. *The Dilemmas of an Upright Man: Max Planck and the Fortunes of German Science.* Cambridge, Mass.: Harvard University Press, 2000. First published 1986.

Moore, Walter. *Schrödinger: Life and Thought.* New York: Cambridge University Press, 2015. First published 1989.

Pais, Abraham. *Niels Bohr's Times in Physics, Philosophy, and Polity.* New York: Oxford University Press, 1991.

*Rozental, Stefan, ed. *Niels Bohr: His Life and Work as Seen by His Friends and Colleagues.* Hoboken, N.J.: Wiley, 1967.

ASTRONOMY

Christianson, Gale E. *Edwin Hubble: Mariner of the Nebulae.* Chicago: University of Chicago Press, 1995.

Douglas, Vibert. *The Life of Arthur Stanley Eddington.* London: Thomas Nelson, 1956.

Ferris, Timothy. *Coming of Age in the Milky Way.* New York: Perennial, 2003. First published 1988.

Johnson, George. *Miss Leavitt's Stars: The Untold Story of the Woman Who Discovered How to Measure the Universe.* New York: Norton, 2005.

Levenson, Thomas. *The Hunt for Vulcan . . . and How Albert Einstein Destroyed a Planet, Discovered Relativity, and Deciphered the Universe.* New York: Random House, 2015.

*Miller, Arthur I. *Empire of the Stars: Obsession, Friendship, and Betrayal in the Quest for Black Holes.* New York: Houghton Mifflin, 2005.

*Singh, Simon. *Big Bang: The Origin of the Universe.* New York: HarperCollins, 2004.

Notes

Prologue

page

xii *building card castles: The Collected Papers of Albert Einstein,* vol. 1, *The Early Years, 1879–1902,* trans. Anna Beck (Princeton: Princeton University Press, 1987), p. xix (hereafter cited as CPAE1). Princeton University Press has been bringing together all of Einstein's papers in a collection that sets the standard for scholarly editions.

"I might not be more skilled": Ernst Straus (Einstein's assistant at Princeton, late 1940s), in *Einstein: A Centenary Volume,* ed. A. P. French (Cambridge, Mass.: Harvard University Press, 1979), p. 31.

xiii *"My boldest dreams": The Collected Papers of Albert Einstein,* vol. 8, *The Berlin Years: Correspondence, 1914–1918,* trans. Ann M. Hentschel (Princeton: Princeton University Press, 1998), p. 160 (hereafter cited as CPAE8).

"the greatest blunder": Did he say it? The first mention was by physicist George Gamow in 1956, the year after Einstein's death. Since Gamow had an ingenious turn of phrase and Einstein never used the term in other correspondence, some historians have suggested that Gamow made it up. I believe Gamow, however. He was highly respected, and his remarks about other colleagues hold up. Most of all, the phrase matches Einstein's tone and sentiments: he was *not* happy about having to add the cosmological constant.

1. *Victorian Childhood*

7 *"The teachers . . . seemed to me":* Philipp Frank, *Einstein: His Life and Times,* rev. ed. (New York: Knopf, 1953), p. 11.

"Einstein, you'll never": CPAE1, p. xx.

"I got used a long time ago": CPAE1, p. 11.

8 *"Beloved sweetheart"*: CPAE1, pp. 11, 12.

9 *"It is nothing short of a miracle"*: Paul Arthur Schilpp, *Albert Einstein: Philosopher-Scientist* (LaSalle, Ill.: Open Court Press, 1949), pp. 16–17.

10 *"director's reprimand for nondiligence"*: CPAE1, p. 27.
"Einstein the eagle": Carl Seelig, *Albert Einstein: A Documentary Biography* (London: Staples Press, 1956), p. 71.

11 *"I would rather not speculate"*: Ibid., p. 11.

12 *"Without you . . . I lack self confidence"*: CPAE1, p. 145.
"Michele has already noticed": CPAE1, p. 152.

14 *"Strenuous intellectual work"*: CPAE1, p. 32.

2. *Coming of Age*

16 *"What a waste"*: CPAE1, p. 152.

17 *"with the humble inquiry"*: CPAE1, p. 151.
"I will soon have graced": CPAE1, p. 163.
"What oppresses me most": CPAE1, p. 123.
"my son Albert": CPAE1, p. 165.

18 *"When I found your letter"*: CPAE1, p. 165.
"My dear dolly!": CPAE1, p. 173. I've very slightly edited the ending.

19 *"a very small [horse-drawn] sledge"*: CPAE1, p. 172.
"How beautiful it was": Walter Isaacson, *Einstein: His Life and Universe* (New York: Simon & Schuster, 2007), p. 64.
"It has really turned out": CPAE1, p. 191.

20 *"Private lessons"*: CPAE1, p. 192.
"is 5 ft 9": Seelig, p. 58.

21 *"We shall remain"*: CPAE1, p. 186.
how much she hated Miss Marić: In July 1900, Einstein told his family he was going to marry Marić. He remembered that "Mama [then] threw herself on the bed, buried her head in the pillow, and cried like a child." Then she said that he was ruining his future, that no decent family would have her, and that "if she gets pregnant you'll really be in a pretty mess" (CPAE1, pp. 141–42). Aside from that, she took it pretty well.
"a married man now": *Albert Einstein–Michele Besso Correspondance, 1903–1955*, trans. and ed. Pierre Speziali (Paris: Hermann, 1972), p. 3.
"I like him a great deal": Albrecht Folsing, *Albert Einstein: A Biography*, trans. and abr. Ewald Osers (New York: Viking, 1997), p. 73.

22 *"become so incomprehensible to each other"*: CPAE1, p. 129.
"When I talked about experiments with clocks": Frank, p. 131.

3. Annus Mirabilis

24 *"a temptation to superficiality"*: Folsing, p. 102.
"The sight of the twinkling stars": Maurice Solovine, *Albert Einstein: Letters to Solovine* (New York: Philosophical Library, 1987), p. 6.

25 *The terms used by early scientists:* Stephen Toulmin and June Goodfield's old classic *The Architecture of Matter* (London: Hutchinson, 1962) traces the underlying concepts back long before the time of Lavoisier, as does Max Jammer's *Concepts of Mass in Classical and Modern Physics* (New York: Dover, 1997). C. E. Perrin's essay "The Chemical Revolution: Shifts in Guiding Assumptions" ("The Chemical Revolution: Essays in Reinterpretation," special issue, *Osiris*, 2nd ser. [1988]: pp. 53–81) is excellent on what happened in Lavoisier's time that made the later jump to a focus on mass so difficult. Charis Anastopoulos's *Particle or Wave: The Evolution of the Concept of Matter in Modern Physics* (Princeton: Princeton University Press, 2008) shows how that modern view developed.

27 *"This should suffice to show"*: CPAE1, p. xviii.

29 *"[Their] religious feeling takes the form"*: Albert Einstein, "The Religious Spirit of Science," in *Ideas and Opinions* (London: Folio Society, 2010), p. 38.

32 *"Perhaps it will prove possible": The Collected Papers of Albert Einstein,* vol. 2, *The Swiss Years: Writings, 1900–1909,* trans. Anna Beck (Princeton: Princeton University Press, 1989), p. 24.

33 *"The idea is amusing": The Collected Papers of Albert Einstein,* vol. 5, *The Swiss Years: Correspondence, 1902–1914,* trans. Anna Beck (Princeton: Princeton University Press, 1995), doc. 28 (hereafter cited as CPAE5).
"Both of us, alas, dead drunk": Dennis Overbye, *Einstein in Love: A Scientific Romance* (New York: Viking, 2000), p. 139.

4. Only the Beginning

34 *When von Laue arrived and made his inquiries:* Seelig, pp. 92–93.

36 *"This unconstruable and unvisualizable dogmatism"*: Folsing, p. 203.

37 *"Perhaps it would be possible"*: CPAE5, p. 20.

5. Glimpsing a Solution

51 *"the happiest thought of my life": The Collected Papers of Albert Einstein,* vol. 7, *The Berlin Years: Writings, 1918–1921,* trans. Alfred Engel (Princeton: Princeton University Press, 2002), p. 31.

52 *"withdraw to the sofa"*: CPAE1, p. xxii.
54 *"With that kind of fame"*: Mileva Einstein-Marić, *In Albert's Shadow: The Life and Letters of Mileva Marić, Einstein's First Wife,* ed. Milan Popovic (Baltimore: Johns Hopkins University Press, 2003), p. 14.
"Isn't it clear": Overbye, p. 185.
55 *"due to my poor memory"*: Folsing, p. 259.
"If it is possible": Seelig, p. 95.

6. Time to Think

57 *"We are on very good terms"*: Peter Galison, Gerald Holton, and Silva S. Schweber, eds., *Einstein for the 21st Century: His Legacy in Science, Art, and Modern Culture* (Princeton: Princeton University Press, 2008), p. 186.
"the size of a visiting card": Seelig, p. 171.
a reassuring nod: Ronald W. Clark, *Einstein: The Life and Times* (New York: Avon, 1971), p. 322.
61 *"the best sounding board"*: Seelig, p. 85.
62 *"I would have loved"*: CPAE5, doc. 300.
63 *"Grossmann, you've got to help me"*: Abraham Pais, *Subtle Is the Lord: The Science and Life of Albert Einstein* (New York: Oxford University Press, 1982), p. 212.

7. Sharpening the Tools

67 *"created a new universe"*: Jeremy Gray, *Worlds out of Nothing: A Course in the History of Geometry in the 19th Century* (London: Springer, 2007), p. 129.
"appear to be paradoxical": Marvin Jay Greenberg, *Euclidean and Non-Euclidean Geometries: Development and History* (New York: W. H. Freeman, 2007), p. 191.
"Grossmann is getting his doctorate": CPAE1, p. 190.
"I have become imbued": Banesh Hoffmann, *Albert Einstein: Creator and Rebel* (New York: Viking, 1972), p. 116.
69 *"exclusively on the gravitation problem"*: CPAE5, p. 324.
"not the kind of vagabond": Seelig, p. 10.
"split into numerous specialties": French, p. 15.
"when I sat on a chair": Hoffmann, p. 117.
"too complicated": Jurgen Neffe, *Einstein: A Biography,* trans. Shelley Frisch (New York: Farrar, Straus and Giroux, 2007), p. 219.
70 *"Compared with this problem"*: Ibid., p. 116.

"Einstein is stuck so deep": Armin Harmann, *The Genesis of Quantum Theory, 1899–1912* (Cambridge, Mass.: MIT Press, 1971), p. 69.

"Never in my life": Hoffmann, p. 116.

"the tail of the lion": CPAE5, doc. 513.

71 *"completely necessary for social reasons"*: Alice Calaprice, ed., *The Ultimate Quotable Einstein* (Princeton: Princeton University Press, 2011), p. 37.

72 *"When I talk to people"*: Folsing, p. 399.

8. *The Greatest Idea*

76 *"Perhaps in another life"*: G. Waldo Dunnington, *Carl Friedrich Gauss: Titan of Science* (1955; repr., New York: Mathematical Association of America, 2004), p. 465.

78 *"after long years of searching"*: John Stachel, *Einstein from "B" to "Z"* (Boston: Birkhäuser, 2002), p. 232.

"the greatest satisfaction": Folsing, p. 369.

"My boldest dreams": Ibid., p. 374.

9. *True or False?*

83 *"My colleagues concerned themselves"*: Pais, p. 235.

84 *"As an older friend"*: Ibid., p. 239.

"so many major hitches": Folsing, p. 317.

90 *"If the Academy won't play ball"*: Ibid., p. 320.

92 *"a 'multitude of the most sophisticated measurements'"*: Ibid., p. 382.

10. *Totality*

93 *a lean, sweaty Englishman:* Eddington didn't mention sweating in his journal, but in May, on the equator off the Congo coast, manipulating heavy equipment outside at midday, anyone is going to sweat.

95 *"This postscript"*: Subrahmanyan Chandrasekhar, *Eddington: The Most Distinguished Astrophysicist of His Time* (Cambridge: Cambridge University Press, 1983), p. 25.

96 *"one person who heads an expedition"*: Quoted in Matthew Stanley, "'An Expedition to Heal the Wounds of War': The 1919 Eclipse and Eddington as Quaker Adventurer," *Isis* 94 (2003): p. 68.

"The lines of latitude": Ibid., p. 64.

98 *"It was impossible to get any work done"*: F. W. Dyson, A. S. Eddington,

and C. Davidson, "A Determination of the Deflection of Light by the Sun's Gravitational Field, from Observations Made at the Total Eclipse of May 29, 1919," *Philosophical Transactions of the Royal Society A,* 20, nos. 571–81 (January 1, 1920), http://rsta.royalsocietypublishing.org/content/roypta/220/571-581/291.full.pdf. This article, the Stanley article cited in the previous note, and an article by Peter Coles ("Einstein, Eddington and the 1919 Eclipse," *Astronomical Society of the Pacific Conference Proceedings,* 252 [2001]: p. 21) are the central sources for the information about Eddington in this chapter.

103 *"Have you by any chance":* Einstein to Paul Ehrenfest, September 12, 1919, in *The Collected Papers of Albert Einstein,* vol. 9, *The Berlin Years: Correspondence, January 1919–April 1920,* trans. Ann M. Hentschel (Princeton: Princeton University Press, 2004), doc. 104 (hereafter cited as CPAE9).

104 *"The whole atmosphere":* Alfred North Whitehead, *Science and the Modern World* (1925; repr., New York: Free Press, 1967), p. 13.
"After a careful study": "Joint eclipse meeting of the Royal Society and the Royal Astronomical Society," *The Observatory* (1919), p. 391.

105 *"We owe it to that great man":* The objector was the physicist Ludwik Silberstein, quoted in *Times* (London), November 7, 1919.
"This is the most important result": Quoted ibid.

11. *Cracks in the Foundation*

113 *"gloomy veil":* G. J. Whitrow, *Einstein: The Man and His Achievement* (London: Dover, 1967), p. 20.
"[an] excellent and truly enjoyable relationship": CPAE8, doc. 56.
"no mental brainstorm": Neffe, p. 102.
"my mother often said": Ibid., p. 103.
"acted upon women as a magnet": Ibid., p. 106. The architect was their close family friend Konrad Wachsmann, who designed the Einsteins' country home.

114 *"The Austrian woman":* Roger Highfield and Paul Carter, *The Private Lives of Albert Einstein* (Boston: Faber and Faber, 1993), p. 208.

115 *"What I admired most":* Einstein to Besso's adult children, March 2, 1955, in *Albert Einstein–Michele Besso Correspondance,* p. 537.
"larger portions of the physical universe": Albert Einstein, "Cosmological Considerations on the General Theory of Relativity," in H. A. Lorentz, A. Einstein, H. Minkowski, and H. Weyl, *The Principle of Relativity: A Collection of Original Memoirs on the Special and General Theory of Relativity* (1923; repr., New York: Dover, 1952), p. 177.

117 *"I have come to the conclusion"*: Ibid., p. 180.
118 *"That term"*: Ibid., p. 188.
"gravely detrimental to the formal beauty": Ibid., p. 193.

12. *Rising Tensions*

125 *"My life is fairly even"*: Eduard A. Tropp, *Alexander A. Friedmann: The Man Who Made the Universe Expand* (Cambridge: Cambridge University Press, 1993), p. 70.
126 *"with excellent organization"*: Ibid., p. 74.
"The distance between our airplanes": Ibid., pp. 75–76.
127 *"what Hindu mythology has to say"*: Helge Kragh, *Cosmology and Controversy: The Historical Development of Two Theories of the Universe* (Princeton: Princeton University Press, 1996), p. 25.
128 *"The results . . . contained in [Friedmann's] work"*: Tropp, p. 169.
"Allow me to present": Ibid., p. 171.
"Science, once our greatest pride": Folsing, p. 524.
129 *"Of all the people"*: Isaacson, p. 307.
130 *"In my previous note"*: Tropp, p. 172.
132 *"There is a wild currency orgy"*: Ibid., p. 187.
133 *"My trip is not going well"*: Ibid., p. 173.
"Everybody was much impressed": Ibid., p. 174.

13. *The Queen of Hearts Is Black*

145 "mais votre physique": A. L. Berger, ed., *The Big Bang and Georges Lemaître: Proceedings of a Symposium in Honour of G. Lemaître, Louvain-la-Neuve, Belgium* (Dordrecht: D. Reidel, 1983), p. 370.
146 *"He didn't seem at all"*: H. Nussbaumer and L. Bieri, *Discovering the Expanding Universe* (Cambridge: Cambridge University Press, 2009), p. 111.
147 *felt terribly uncomfortable:* The experiment was conducted by Jerome S. Bruner and Leo Postman, and their results were published in "On the Perception of Incongruity: A Paradigm," *Journal of Personality* 18 (1949): pp. 206–23. "Perhaps the most central finding," they wrote, "is that the recognition threshold for the incongruous playing cards (those with suit and color reversed) is significantly higher than the threshold for normal cards. While normal cards on the average were recognized correctly—here defined as a correct response followed by a second correct response—at 28 milliseconds, the incongruous cards required 114 milliseconds." No wonder Einstein held on to his mistake as long as he did.

147 *"to see wounded men fall"*: Gale E. Christianson, *Edwin Hubble: Mariner of the Nebulae* (Chicago: University of Chicago Press, 1995), p. 108.

153 *"[I] have netted nine novae"*: Robert W. Smith, *The Expanding Universe: Astronomy's "Great Debate," 1901–1931* (Cambridge: Cambridge University Press, 1982), p. 114.

155 *"It means nothing"*: Arthur I. Miller, *Einstein, Picasso: Space, Time, and the Beauty That Causes Havoc* (New York: Basic Books, 2001), p. 235.

156 *"muscled his way in"*: Christianson, p. 206.
"New observations by Hubble and Humason": Ibid., p. 210.

157 *"It is remarkable"*: Daryl Janzen, "Einstein's Cosmological Considerations" (unpublished paper, University of Saskatchewan, Saskatoon, February 13, 2014), http://arxiv.org/pdf/1402.3212.pdf, pp. 20–21.
"When life is full of trouble": Christianson, p. 211.

14. *Finally at Ease*

158 *"Since I introduced this term"*: Kragh, p. 54.
"This is the most beautiful": Timothy Ferris, *Coming of Age in the Milky Way* (1988; repr., New York: Perennial, 2003), p. 212.
"some very interesting things": Berger, p. 376.

159 "Ah, très joli": Ibid., p. 376.

162 *"you have proved"*: "Dark Side of Einstein Emerges in His Letters," *New York Times*, November 1996.
"I must love someone": CPAE5, doc. 389.

163 *Was it really flying snakes:* Dorothy Michelson Livingston, *The Master of Light* (New York: Scribner's, 1973), p. 291, cited in Denis Brian, *Einstein: A Life* (New York: Wiley, 1996), p. 12.
"exceptionally kindhearted": Neffe, p. 102.

164 *"ten years younger"*: Isaacson, p. 361.
"As Genesis suggested it": Berger, p. 395.

165 *"The evolution of the universe"*: Simon Singh, *Big Bang: The Origin of the Universe* (New York: HarperCollins, 2004), p. 159.
"the most beautiful": Ferris, p. 212.
accurate equations that are "smarter" than the people who devised them: The great example here is Paul Dirac's equation describing the electron, which he published in 1928. An equation such as $x^2=25$ has two solutions: $x=5$, or $x=-5$. Dirac's equation also had two possible solutions: one for negatively charged electrons—which were all the electrons then known—yet another for positively charged electrons, although such "positive" electrons were utterly unimagined. Four years later, Carl Anderson at

Caltech discovered them, and this is what prompted Dirac to observe, "My equation is smarter than I am." How could this happen? See, for example, Frank Wilczek, "The Dirac Equation," in *It Must Be Beautiful: Great Equations of Modern Science,* ed. Graham Farmelo (London: Granta, 2002), pp. 132–61. See also Eugene P. Wigner's much-anthologized 1960 paper, "The Unreasonable Effectiveness of Mathematics in the Natural Sciences."

15. Crushing the Upstart

172 *"quite the most incredible":* Michael Hiltzik, *Big Science: Ernest Lawrence and the Invention That Launched the Military-Industrial Complex* (New York: Simon & Schuster, 2015), p. 18.

176 *"The weakness of the theory": The Collected Papers of Albert Einstein,* vol. 6, *The Berlin Years: Writings, 1914–1917,* trans. Alfred Engel (Princeton: Princeton University Press, 1997), p. 396.

helpful to talk about the probabilities: The probabilities had to go in because it was the only way to deduce Planck's already well-known radiation law. But that is what Einstein was convinced was just a stopgap. Bohr, however, welcomed the probability approach, because in his atomic theory, transition processes could *never* be understood classically.

177 *"The real joke":* Folsing, p. 393.

180 *"It was almost three o'clock":* Jagdish Mehra, ed., *The Golden Age of Theoretical Physics: Selected Essays* (London: World Scientific, 2001), pp. 651–52.

181 *"The Heisenberg-Born concepts":* Max Born, *The Born-Einstein Letters, 1916–1955: Friendship, Politics and Physics in Uncertain Times,* trans. Irene Born (1971; repr., London: Macmillan, 2005), p. 86.

"Quantum mechanics is certainly imposing": Ibid., p. 88.

"a big quantum egg": Folsing, p. 566.

16. Uncertainty of the Modern Age

184 *"convinced of the causal dependence":* Pais, p. 467.

being an atheist was presumptuous: In 1936 Einstein wrote, "Everyone who is seriously involved in the pursuit of science becomes convinced that a spirit is manifest in the laws of the Universe—a spirit vastly superior to that of man" (Calaprice, p. 152). In 1941: "The fanatical atheists are like slaves who are still feeling the weight of their chains . . . They are creatures who—in their grudge against traditional religion as the 'opium of the masses'—cannot hear the music of the spheres" (Isaacson, p. 390).

184 *"the guiding principle"*: Einstein, "The Religious Spirit of Science," p. 38.

185 *"Einstein was not at all satisfied"*: Werner Heisenberg, *Encounters with Einstein: And Other Essays on People, Places, and Particles* (Princeton: Princeton University Press, 1983), pp. 113–14.

186 *"A good joke"*: Frank, p. 216.

"a veritable witches' multiplication table": Folsing, p. 580.

187 *Schrödinger had worked out his equation:* Schrödinger's wife was always helpful when it came to advancing quantum physics, and a few months later she provided twin sisters to aid her husband's concentration. "Nirvana is a state of pure blissful knowledge," Schrödinger wrote. ". . . It has nothing to do with the individual" (Walter Moore, *Schrödinger: Life and Thought* [1989; repr. New York: Cambridge University Press, 2015], p. 223).

189 *"The more I think"*: Ian Stewart, *Why Beauty Is Truth* (New York: Basic Books, 2007), p. 209.

"I got the idea of investigating": Stefan Rozental, ed., *Niels Bohr: His Life and Work as Seen by His Friends and Colleagues* (Hoboken, N.J.: Wiley, 1967), p. 106.

17. *Arguing with the Dane*

192 *"an inner voice"*: *Born*, p. 88.

193 *"Not often in life"*: Calaprice, p. 61.

196 *"In the course of the day"*: Heisenberg, p. 116.

"a perpetual motion machine": Folsing, p. 589.

197 *"I believe that the limitation"*: Max Planck award ceremony, June 28, 1929, in Calaprice, p. 172.

"Carry on!": French, p. 15.

198 *now almost universally accepted:* The formal experimental proof only came four years later, in 1933, with the American physicist Kenneth Bainbridge using a sensitive mass spectrometer. But few physicists really needed that: Einstein's field equations that led to the light-bending prediction depended on $E=mc^2$, and when Eddington had so spectacularly confirmed the theory in 1919, he had indirectly confirmed $E=mc^2$.

201 *"[Bohr] was extremely unhappy"*: Pais, p. 446.

202 *"he would not give up"*: Rozental, p. 103.

203 *"we could now be sure"*: Heisenberg, p. 116.

18. Dispersions

212 *"spectators and actors"*: David Cassidy, *Uncertainty: The Life and Science of Werner Heisenberg* (New York: W. H. Freeman, 1992), p. 545.
214 *"German men and women!"*: Mordecai Schreiber, *Explaining the Holocaust: How and Why It Happened* (Eugene, Ore.: Cascade Books, 2015), p. 57.
215 *"Look around you"*: Frank, p. 226.

19. Isolation in Princeton

216 *"a quaint and ceremonious village"*: Einstein to Queen Elizabeth of Belgium, November 20, 1933, quoted in Calaprice, p. 25.
217 *"so upset by my illness"*: Antonina Vallentin, *The Drama of Albert Einstein* (New York: Doubleday, 1954), p. 240.
"settled down splendidly": Born, p. 128.
218 *"I still do not believe"*: Folsing, p. 704.
220 *"You are a smart boy, Einstein"*: Pais, p. 44.
222 *"to have one wife at Oxford"*: Moore, p. 298.
"You are my closest brother": Ibid., p. 426.
224 *"no self-respecting person"*: Isaacson, p. 431.
225 *"each time he does that"*: Folsing, p. 127.
"it would be better not to work": Ibid., p. 695.
"a petrified object": Born, p. 178.
226 *"Bohr was profoundly unhappy"*: Folsing, p. 705.

20. The End

227 *"I know what's wrong"*: Recalled by Einstein's assistant Ernst Straus at a 1955 memorial talk, quoted in Calaprice, p. 192.
228 *"I miss her more"*: Interview with Hanna Loewy, an old family friend, 1991, quoted in Calaprice, p. 32.
"Einstein hardly referred": "A Genius Finds Inspiration in the Music of Another," *New York Times*, January 31, 2006.
229 *"in an airship"*: Isaacson, p. 511.
"The foundation of our friendship": Hoffmann, p. 257.
230 *"To think with fear"*: Ibid., p. 261.
"It is tasteless to prolong": Pais, p. 477.
"He joked with me": Max Born, *My Life: Recollections of a Nobel Laureate* (New York: Scribner's, 1978), p. 309.

230 *"If only I had more mathematics":* Peter Michelmore, *Einstein: Profile of the Man* (New York: Dodd, Mead, 1962), p. 261.

Epilogue

232 *"was always in a good mood":* French, p. 13.

234 *starting in the 1990s:* Two teams, announcing their results within weeks of each other in 1998, were awarded the 2011 Nobel Prize in Physics for this work. To find that the universe was expanding ever faster was, as Saul Perlmutter, head of the California team, told National Public Radio interviewer Terry Gross, "a little bit like throwing [an] apple up in the air and seeing it blast off into space" (*Fresh Air,* NPR, November 14, 2011).

Would this have troubled Einstein? Probably not, for these findings didn't necessarily require the conceptual approach of quantum mechanics; they could still, one might think, be explained in clear, causal terms.

In April 1917, Einstein had written to the Göttingen mathematician Felix Klein, "I don't doubt that sooner or later [my theory] will have to give way to another that differs from it fundamentally, for reasons that today we cannot even imagine.

"I believe that this process of deepening theory has no limit."

Appendix

238 *"Einstein's presentation":* Constance Reid, *Hilbert* (New York: Springer, 1996; orig. 1970), p. 112.

"objects of our perception": H. Minkowski, "Space and Time," in H. A. Lorentz, A. Einstein, H. Minkowski, and H. Weyl, *The Principle of Relativity: A Collection of Original Memoirs on the Special and General Theory of Relativity* (1923; repr., New York: Dover, 1952), p. 76.

240 *"this most valiant piece":* Ibid., p. 76

242 *"superfluous erudition":* Pais, p. 151.

247 *"Where wast thou":* Book of Job, 38: 4-6.

Index

Page locators in *italics* indicate figures and photographs.